The River Cottage

Bees & Honey Handbook

The River Cottage Bees & Honey Handbook

by Steve Minshall

& Rachel de Thample

with an introduction by

Hugh Fearnley-Whittingstall

rivercottage.net

BLOOMSBURY PUBLISHING

LONDON • OXFORD • NEW YORK • NEW DELHI • SYDNEY

For Chris

with thanks for her love, support and encouragement over many years ... Steve

& Theodora

who will be buzzing with delight that this book finally made it... Rachel

BLOOMSBURY PUBLISHING
Bloomsbury Publishing Plc
50 Bedford Square, London, WC1B 3DP, UK
Bloomsbury Publishing Ireland Limited, 29 Earlsfort Terrace, Dublin 2, D02 AY28, Ireland

BLOOMSBURY, BLOOMSBURY PUBLISHING and the Diana logo are trademarks of Bloomsbury Publishing Plc

First published in Great Britain 2025

A catalogue record for this book is available from the British Library

ISBN: HB: 978-1-4088-7355-7

2 4 6 8 10 9 7 5 3 1

Project Editor: Janet Illsley
Designer: Will Webb
Photographers: Steve Minshall and Ali Allen
Indexer: Hilary Bird

Printed in Slovenia by DZS Grafik

To find out more about our authors and books visit www.bloomsbury.com and sign up for our newsletters

For product safety related questions contact productsafety@bloomsbury.com

Contents

The Importance of Bees 8

Making Your Space More Bee Friendly 34

Understanding Honeybees 56

Keeping Honeybees 74

The Beekeeper's Year 126

The Magic of Honey, Beeswax and Propolis 138

Rachel's Recipes 166

Useful Things 248

This book is foremost a celebration of the remarkable work of the honeybee. And it's a warning too, reminding us that bees are not just nice creatures to have around, they are vital to our environment, to the food we eat, to the whole way in which life functions on our planet. As Rachel and Steve put it so aptly in their excellent introduction, bees are 'one of the strongest threads in nature's complex web'. If we don't take care of them, things really will fall apart.

Rachel and Steve's advice is practical and inspiring for all, emphasising that you can enjoy and celebrate bees without ever having to don a veil and a pair of stout gloves, let alone pick up a smoker or a bee brush. Observation is one of the key ways we can interact with the natural world, and bees offer the perfect opportunity for this. Read the first part of this book and you'll never look at a small buzzing insect in the same way again. In fact, you'll probably be settling down outdoors on a sunny summer's day to see how many different bee species you can spot. The more time we spend with bees, the more we will care about them, and the more invested we will be in their survival.

Secondly, if you've got a garden or a balcony, fill it with bee-friendly plants (mostly in the blue-purple-pink spectrum, if that works for you, as this is apparently bees' favourite colour range!). The list of such plants is extensive and includes many inexpensive, easy-to-grow cottage garden favourites like lavender and borage, rosemary, foxgloves and wallflowers. Limit your mowing: letting your garden get a bit wild and untidy is brilliant for bees who will relish an abundance of blooms, including 'weeds', and a diversity of plants overall. If you don't have a garden, seek out places with flowering plants like bramble, buddleia or deadnettle. The joy of seeing pollinators busily at work on these and other species is a mood-booster for me – I hope it will lift your spirits too.

If you already keep bees, or would like to, you will find extensive information and advice in these pages. But this is very much a book for all of us, whether or not we host these incredible creatures in our own hives. You might want to readdress your relationship with honey, choosing only what is produced locally to you and raw or minimally processed. It will cost more than mass-produced, blended honey, but it will be more distinctive in flavour and offer you more goodness, too. Or you might, especially if you are vegan, choose not to eat honey at all but simply to celebrate the complex creatures who make it as part of their fascinating life cycle.

There are almost 300 individual species of bee in Britain alone. But the domesticated honeybee represents a tiny fraction of this total, with just one species, the Buckfast bee, predominating. So, while honey-producing beekeepers have almost certainly turned to their art out of a wish to support and enhance nature, we are starting to realise that the picture is a little more complex. A great number of productive honeybee hives full of only one type of bee is actually not good news for biodiversity overall.

At River Cottage, we still harvest honey. But we are starting to move away from conventional beekeeping, with its neat slatted National hives, extractors and spacesuit-like protective gear. We also have Top bar hives and Warré hives, and Steve has years of experience to guide us through the features that will help beginners in beekeeping make the choices that work for them.

And as our knowledge of bees deepens, we are exploring more natural beekeeping techniques. This includes the Freedom hive, pioneered by our friend Matt Somerville, which is essentially a hollowed-out log that provides a home for wild bees, from which we don't collect any honey at all. I have a couple of these at home and they bring me so much joy. By fostering and encouraging bees to simply do what they do, in a well-insulated log hive that simulates the choice they would make in the wild, we are nurturing the resilience and genetic diversity of self-determining wild (or semi-wild) bee populations. We are going some way to turn the tide of destruction that has been unleashed on them.

Bees are under immense threat from all sides: from prolific and irresponsible pesticide use, from monoculture crops that don't allow them the varied diet they need, and even from invasive species like the Yellow-legged (or Asian) hornet… the list goes on. But we have the information and, I hope, the will to change that.

Rachel's wonderful 'less is more' attitude to using honey in the kitchen honours and respects what these little workers do, and how hard they must labour to do it. If we are to eschew cheap, hyper-filtered, heated, multi-source honeys in favour of the real stuff, we have to use it more wisely too. It's something I've been on a personal journey with too – I adore honey, but the knowledge of how much natural goodness comes with a spoonful of very good local raw honey, in comparison to a jugful of the mediocre, means that, like Rachel, I get more out of it even as I consume less of it.

Britain is among the most nature-depleted countries on the planet and bees are all-important to us. They are, in all their fabulous variety, a keystone species, both reflecting and promoting biodiversity – but only if we let them. Keeping these fascinating insects in our minds – and doing all we can to protect and attract them – is a crucial step in the right direction. There's a whole spectrum of ways you can get involved with bees, from watching them to working with them, and this little handbook is the best possible place to start.

Hugh Fearnley-Whittingstall, East Devon, October 2024

The Importance of Bees

Worker honeybees returning to their hive

Life without bees would be not only be sad and silent, it would also be less delicious and nutritionally diverse. One of the reasons why humans have evolved such a fascination with bees is the lure of the concentrated nectar they produce. Our brains are hard-wired to love them, as honey gives us an instant buzz. It is, after all, the most potent brain-food, as well as nature's most energy-dense source of calories.

Of course, the human obsession with bees is much deeper. Our winged friends have helped to shape philosophical, economic and political landscapes. Nestling closer to the ground, attuning your ears to the unique hums of different species and seeing the intricate design of pollinating insects feeds a deeper hunger in which modern life has made us deficient. But the most compelling reason to love bees – all bees, not just honeybees – is that they provide pivotal pollination services and are one of the strongest threads in nature's complex web.

Sadly, however, we are losing insect species at an alarming rate and bees are top of the tree. Around a quarter of bee species haven't been seen by scientists in the last 30 years. In the UK, the bumble bee is one of the vanishing species and many other bee species are in a precarious state. Colony collapse in Europe and across the globe is worryingly high. But there are simple ways in which we can help bees. Two of the most important things you can do to is to plant a diverse range of flowers they can feast from throughout the year (see pp.52–5), and support chemical-free farming.

Bees made headlines and stirred the public to jump to action after Colony Collapse Disorder (CCD) was coined in 2006, following dramatic disappearances of honeybees. Entire populations on farms in America suddenly vanished. Just before the turn of the twenty-first century, apple orchards in China's Maoxian County, Szechuan Province went silent as the spring blossom set with no insects to spread the pollen to set the fruit. Farmers had to be employed in place of bees to hand-pollinate flowers using brushes and cigarette filters attached to chopsticks. The use of pesticides, a decline in habitats and lack of forage are key culprits that keep surfacing when scientists explore the plight of our pollinators.

There are a staggering 20,000 different bee species worldwide (with around 270 different types of bees in the UK alone) and they all work together to help feed us. Nearly three-quarters of crops that produce fruits and seeds for human consumption depend on bees. And wild bees are as important, if not more so, than honeybees in making harvests happen. Well-intended efforts to keep bees in gardens has led to unintended consequences, with some wild bee species now struggling to find enough forage.

The motivation of most prospective beekeepers is to help 'save the bees'. While setting up a hive is a wonderful way to learn more about bees, collect honey and engage with our winged friends in an intimate way, it's not the solution to the problem. In fact, since Colony Collapse Disorder was reported, the number of hobbyist beekeepers in the UK has soared to such heights that, in some areas, the concentration of bees is putting pressure on other wildlife, if not the honeybees themselves. In some areas of London, for example, as many as 400 hives are packed into a square kilometre; a healthy volume of hives in such a dense urban space is seven.

At River Cottage, we're moving away from the traditional way of beekeeping in favour of creating habitats to help secure the future of wild honeybees, which are in serious decline. The benefit is a more natural home for honeybees with minimal work for us. We're edging away from beekeeping in favour of an exciting approach and we've already seen a hive of wild honeybee activity at Park Farm in our Freedom hive, made out of a hollowed-out log.

The key aim of this book is to explore sustainable paths for enjoying honey and supporting our precious pollinators. Throughout the book, we look at a range of options for you to protect and engage with both honeybees and the wider world of pollinators, from managing a modern hive in the most sustainable way, to ideas on how you can create habitats and foraging corridors for bees. We also explore alternative ways to engage in beekeeping, such as log hives for wild bees.

This might be a relief to some of you who are keen to get involved with bees but have uncertainties regarding time commitments and space. There are so many brilliant ways to be a pollinator protector. And there are ways to produce jars of honey that don't involve having your own hive, most notably community-based hive sharing ventures. If you're in an area that's not overpopulated and you have the time and space, creating a habitat for bees is brilliant on so many levels. This book explores sustainable options and avenues that put all bees and the wider spectrum of pollinators at the forefront.

We hope this book opens your eyes to just how deeply interconnected we are to nature. In our busy lives, with screens and deadlines clouding our vision, it's so easy to lose sight of what really matters. The current imbalanced bee population is a rather sharp echo of a wider equilibrium that needs readjusting. Often unknowingly, we take so much for granted, and tapping into the wider world of bees resets our focus and makes us value things – like a jar of honey – more. Once you realise just how much effort goes into producing just one teaspoonful of honey, you'll see why it was once prized over gold.

The role of bees as pollinators

Pollen is a sex cell of plants, and insects such as bees are essential for moving pollen from the male structures of flowers (called anthers) to female structures (the stigmas) of the same plant species. A flower needs to be pollinated before it can bear fruit. Think of cherry or apple blossom. If these male sex cells are not moved to its female counterpart, we wouldn't have cherries or apples.

Eighty per cent of global food crops depend on animal pollination, with bees of all species the key insects providing this service. Homing in on the economic benefit in the UK specifically related to honeybees, the crop value benefitted by their pollination equates to an annual estimate of £200 million. Zoom in to a single hive in the UK, and one colony of honeybees can help to pollinate £600 worth of crops each year.

In places where there is a serious decline in bees, humans need to pollinate crops by hand, but of course this takes much longer. Without the speedy work of bees, foods that require pollination would become significantly more expensive. It is estimated that the price of apples alone would double in the absence of bees. Approximately 300,000 species of flowering plants worldwide require insect pollination, so imagine how much our food system would change if humans had to take over the reproduction of these plants?

Food sources that bees help perpetuate are not just for human consumption but also for wider scale biodiversity, providing food, shelter and other resources to mammals, birds and other insects. The nectar and pollen from hawthorn blossom, for instance, is an important forage for bees but equally, the fruit produced from their activity provides food for many wild bird species.

Pollination is also vital for the spread and longevity of rare habitats, such as heathland, which offers a breadth of biodiversity and supports many rare species, aside from its cultural and economic importance.

The evolution of bees

If you've ever had difficulty distinguishing a honeybee from a wasp, you're forgiven. In fact, you're wise to their evolutionary connection. In the Cretaceous era that started 145 million years ago, plants evolved and started to produce flowers, which protected their seeds and made reproduction faster. Flowering plants are known as angiosperms. Uniquely, they advertise the plant's reproductive parts with a protruding pollen-producing stamen and female carpels, housing seeds which ripen into fruit. Insect-hunting wasps were lured by the nectar of these new flowering plants and in doing so, migrated to a different diet. In the process, they also accidentally started to spread pollen from plant to plant, which increased the proliferation of flowering plants. The change in diet of these wasps gradually shape-shifted these winged insects into the bees we know and love today.

Bees have inherited several important features from their predecessors, including a constricted narrow waist, a stinger and an instinct to nest. Their slim waist is an important feature as it gives their abdomen flexibility. Originally, this was useful for stinging insect prey held in the wasp's claws, but bees utilise their agility to deposit eggs deep into the cells of the honeycomb. The sting was once used for attacking and killing prey, but bees employ it as a defence mechanism against thieves plundering honey and brood.

Both wasps and bees pollinate crops, which is why they're considered to be vital species on the planet, but beyond their imperative work as a bigger part of nature, it's their inner world that makes them one of the most complex of all insect societies, especially honeybee colonies. Unlike wasps and other species of bee, honeybee colonies are effectively immortal. As individual bees die, they are quickly replaced – workers week by week and month by month, and queens every few years, but the colony can go on indefinitely. Only honeybees and bumble bees live in colonies and belong to a group called social bees. The majority of other species do not live in colonies and are classified as solitary bees.

While honeybees are unique in their perpetuity, you can trace an evolutionary path from simpler nesting strategies. Many species of bee, or indeed wasp, build their nests by creating tunnels in the soil or burrows in dead wood, or even utilising hollow plant stems for shelter. These tunnels are divided into cells, each with an egg nestled inside and each with a store of food to feed the larvae – insect remains for the wasps and a mixture of pollen and nectar for the bees. Each nest is constructed solely by a lone female working on her own; other nests may be constructed nearby.

Wild bees to 'kept' bees

Humans have long appreciated the vital importance of bees to our survival through their pollination services. And, by closely observing how bees survive in the wild, they have evolved the role of keeper over time. The gradual shift from wild honeybees to domesticated bees developed hand in hand with the growth of agriculture. Shifting from wild to cultivated crops, we realised how important it was to enlist bees to help boost food production. Early farmers recognised that they could increase crop production by having bees on site. Not only would they help pollinate the crops, but they'd also create another food for harvest: honey. But now we've reached a bit of a problem as wild bee populations are now under threat.

Stone Age men cottoned on to the sweet delights of wild honey from bees nesting in sky-reaching tree hollows or virtually inaccessible earthen caves. It was man's first true taste of sweetness and there was no quest too daunting to obtain honey. The earliest record of honey gathering can be found near Valencia in Spain in the Araña Cave (Cave of the Spider) which is thought to date back 7,000 to 15,000 years. Venture inside and etched on the stone interior, in earthen paints, you'll discover the story of Spanish Stone Age man perched perilously high on a rope ladder, reaching longingly towards a bee's nest with one hand, while holding a basket for honeycomb collection in the other. In another cave, at Barranc Fondo, also in Valencia, a scene shows honey hunters working as a group – four staged on a ladder with a posse of accomplices below, eagerly waiting to catch the forthcoming honeycomb.

The rituals of wild honey harvesting are still observed in the foothills of the Himalayas in central Nepal. And there are even urban spots for wild honey foraging, including parts of Berlin, especially Grunewald. In the area around Nuremberg there are still numerous references to honey hunting traditions, and its abundant flow has inspired traditional gingerbread recipes.

The transition from wild to kept bees is thought to have been accidental, triggered by swarms of honeybees nesting in woven baskets or baked-mud water pots left lying around early agricultural settlements along the Nile Delta plain in Egypt. A fertile playground blossomed in the rich silt alongside the river, abundant in bee-loving chamomile, clover, cornflowers, flax and rose.

The Egyptians were quick to harness the benefits of fashioning homes for bees that were far more accessible than the otherwise precarious art of foraging for wild honey. They began creating horizontally staked cylindrical hives made from interwoven twigs and reeds, which would have been covered in sun-dried mud from the Nile or hand-dug clay. At each end was a mud disc with an open flight window for the bees to enter and exit their new dwelling.

The Egyptians quickly found ways to manage the hives, learning how to divide the colony of settled bees by removing panels of comb containing eggs and larvae, thereby allowing honey production to expand. By 1500 BC, beekeeping in Egypt had blossomed into a vast scale.

Beekeeping continued to evolve and prosper, with major advances made by the Ancient Greeks and Romans. Aristotle, a beekeeper himself, offers a detailed picture of progress in *Historia Animalium IX*, written between 344 and 342 BC, where he notes a hive of the time could yield anything from a meagre 3 litres of honey to a prosperous 10 litres.

While the Greeks increased production, the Romans evolved hive design. Roman scholar Varro, 116–27 BC, who also acted as Julius Caesar's librarian, detailed a range of hives in his writings, from round wicker constructions coated with cow dung to rectangular hives measuring 100 x 30cm made from fennel stalks, known as ferula hives, which originated in Sicily. Advancing the hive was motivated by profit. Two brothers by the name of Veianius established an apiary in a garden of thyme, which was so prolific they sequestered the equivalent of £15,000 each year through honey sales.

Throughout the Middle Ages farmers experimented with different styles of beekeeping, some mimicking wild instincts, allowing bees to nest high up in trees in clusters known as bee forests. A solution to reduce harvesting time was to cut the trees down to stump-size so they could be accessed without ladders. Tree stumps housing active colonies inside were almost like totem poles which inspired intricate carvings and designs. You can see surviving examples at the Jan Dzierzon Museum in Kluczbork, Poland, where there is a permanent exhibition on the history of apiculture and Dzierzon's role as the designer of the first successful movable-frame bee hive.

Over the years, beekeepers from all corners of the globe sought to evolve methods of beekeeping as natural as possible for the bees, but also easily accessible and maintainable for mankind. American-born Reverend L.L. Langstroth (1810–95) is credited with the invention of the first fully functioning movable-frame hive, which allowed ample space for bees to move within the hive while giving the beekeeper easy access to inspect the bees and collect the honey.

It is rumoured that Langstroth's design was inspired by a discarded champagne case. The wooden honey storage supers of a Langstroth hive, indeed, are very similar in size and structure. What's brilliant about their construct is that they are reusable, which is why Langstroth hives are still popular. An evolution from the Langstroth is the National Standard hive which is one of the most used in the UK today. It includes an open-mesh floor which allows monitoring of varroa infestation, deeper brood boxes for the queen to lay eggs, and shallower super boxes.

At River Cottage, we've been working with new pioneers who are creating exciting ways to create bee habitats in your garden with minimal interruption for the bees and less work for us. Bee expert Matt Somerville has created a Freedom hive for us at Park Farm (aka River Cottage HQ) in Devon, which mimics the wild habitats, such as tree hollows high from the ground, that bees naturally seek. Such habitats help secure the future of wild honeybees. Wild bees are shown to be more genetically diverse and more environmentally agile, which means that they are more resilient to climate challenges as well as pests and disease.

View of the natural comb bees create inside a Warré hive

1

2

Symbolism of the honeybee

For millennia, in Africa, Asia, Europe and America, bees and their products have inspired the creation of myths. The Greeks even had a god of beekeeping, and a further god of bees and honey. Beekeeping and honey collection has been depicted in ancient cave paintings and Egyptian hieroglyphs. Once tuned in to seeing these images, it's easy to spot them in historic and modern sites – cathedrals can be a rich source. Cities, businesses and families have used hives and bees as heraldic symbols within their crests in order to be associated with their perceived desirable traits. These characteristics are wide ranging, from arduous and dedicated work to sweetness and virtue.

Interesting examples are:

- The powerful Barberini family in seventeenth century Rome used bees in their family crest to symbolise dedication and hard work, while associating themselves with the historic concept of the church as a bee hive, and the clergy as laborious bees. A great example is incorporated in the boat-shaped fountain at the base of the Spanish Steps in Rome, but these crests can be seen throughout Rome, particularly as Pope Urban VIII was from the Barberini family (pic 1).

- In England, the original Lloyds bank logo depicted a skep and bees, to symbolise thrift and industry. There is a fine example on the facade of Lloyds Okehampton branch, Devon (pic 2).

- Depictions of bees and their colonies are not uncommon in cathedrals across Europe. There is a fascinating carving of a bear ravaging a colony of bees in the misericords at Toledo Cathedral, Spain, completed in the fifteenth century (pic 3).

- The City of Manchester's coat of arms depicts seven bees on a globe, to highlight this industrious city's products being exported across the world. Recently, the bee has consistently been used in Manchester to signify solidarity during troubled times.

Nowadays it's easy to find bees depicted on an array of household and decor items, apparently with the intention to highlight the owner's affinity with nature. As a beekeeper, you will gradually find your house filling with these items as gifts from friends and family – resistance is futile!

Bees in the British Isles

The British Isles buzz with an estimated 270 different species of bee (contributing to the 20,000 global tally). The majority of British bees are solitary species. It's an extraordinary feat to learn how to identify the unique characteristics of each, but an illuminating and meditative experience trying.

There are few greater joys than bee watching on a sunny day. It is deeply satisfying when you can spot the subtle (and sometimes more starkly contrasting) differences between species, be it the arrangement of fluffy (or not so downy) bands, the colour of the tail, the pitch of the buzz, or their flower preferences and habitats. The descriptions below give you an insight into the uniqueness of more common individual species and provide a fun guide to take out into the wild for a day of Bee ID.

Honeybees

Archaeological evidence suggests that honeybees have been present here for at least 4,000 years, entering from southern Europe after the last Ice Age. The bee of that period was a dark bee which has recently resurfaced in the British Isles after it was thought to have been wiped out by a deadly mite, called acarine. The mite plagued honeybees in the early twentieth century by attacking the trachea of these dark bees in what is now known as the Isle of Wight disease.

Following their demise, Karl Kehrle, a German Benedictine monk residing in Buckfast Abbey in Devon, known as Brother Adam, bred a new honeybee that was resistant to the tracheal mites, namely the Buckfast bee. It is the most dominant species in Britain, indeed many suggest it is our only species of honeybee but the recent re-emergence of the black bee and newly evolving species are diversifying honeybee ecology, making it an exciting moment in British honeybee history.

When you source bees, Buckfast and Dark British bees are common species you'll come across. The most important bit of advice is to go for a species that thrives in your local area. Your local beekeeping association or local beekeepers will be able to advise you and help you source your bees, but here is a guide to when, where and how to spot the different species.

Dark British bee

When you'll see them out foraging April–September

Where The rare black honeybee which was thought to have been wiped out by a strain of Spanish flu in 1919 was rediscovered in the rafters of a church in Northumberland, where they have been breeding for nearly 100 years.

How to identify Also known as the British black bee, this species has an abdomen that is nearly purely black.

Buckfast bee

Buckfast bee

When you'll see them out foraging April–November

Where Known as the beekeeper's bee, this is the most common type of honeybee in Britain and can be found in most people's gardens, in veg patches, local parks or any area with a proliferation of bee-friendly flowers.

How to identify Buckfast bees have a slim, sandy thorax and black abdomen with golden-amber bands.

Bumble bees

While honeybees have a clear distinction between thorax and abdomen, the body of a bumble bee appears more uniform. Honeybees also have two clear sets of wings: a larger set in the front and a smaller set behind. The other main distinction is honey production. Bumble bees do produce honey but only for self-consumption in their nests. Honeybees, on the contrary, overproduce honey, creating more than they need.

There are 24 different species of bumble bee in the UK, all wild, and it's these fluffy, rounded bees that many of us see zooming around the flowers in local parks, the countryside and back gardens. Bumble bees are a social species, nesting in colonies ranging from a few dozen to several hundred bees. A select few of the

Red-tailed bumble bee

White-tailed bumble bee

different varieties are described below. It's hugely enjoyable to do a bit of bee spotting on a warm spring or hot summer day. You'll be surprised how large some of the early season bumble bees are. These are frequently the queen that is restarting the annual life cycle, i.e. gathering food while laying and rearing a colony.

Red-tailed bumble bee

When you'll see them out foraging April–November
Where These bees do well in a variety of habits including woodland, urban sites, gardens and wildflower-rich grassland; anywhere they can find thistles, bird's-foot trefoil, buddleia and their other favourite flowers.
How to identify Female red-tailed bumble bees are jet black with a bright red or red-orange tail, while males have a yellow-haired head and collar, and a weak yellow midriff band.

White-tailed bumble bee

When you'll see them out foraging March–November
Where White-tailed bumble bees can be found almost anywhere, feeding on flowers ranging from thistles and buddleia to brambles and scabious.
How to identify These bees have a bright yellow collar, a yellow abdomen band

and a bright white tail. They look very similar to buff-tailed bumble bees, which have a browner collar and an orange-tinted tail, and early bumble bees, which are much smaller, with a bright orange or yellow-orange tail.

Buff-tailed bumble bees

When you'll see them out foraging February–late October (or early winter if it's still warm)
Where Cities and rural areas along the south of England, and some parts of Wales and the Midlands. They love feasting on mahonia when it flowers late in the year. They are bred commercially for pollination worldwide but these species are invasive to native bee species elsewhere (Argentina, Chile, Japan and Tasmania).
How to identify The queens are actually the only ones with a buff tail (a fluffy yellow bottom). Workers and drones look more like white-tailed bumbles but they have a small buff stripe just below their central black band.

Early bumble bee

When you'll see them out foraging February–October
Where Geographically, you'll find early bumble bees throughout Britain and Ireland, and widely distributed across mainland Europe. They love foraging

Early bumble bee

Common carder bee

brambles and other wild, overgrown spaces, coastal marshes and brown field sites.
How to identify Smaller than most bumble bees, early bumble bees are also rounder and fluffier than other species. They have a reddish tail, which is much paler and more orange than a red-tailed bumble bee, and a yellow abdominal stripe. The males are more extensively yellow and have very obvious yellow facial fuzz.

Common carder bee

When you'll see them out foraging March–November
Where These bees are very common and found everywhere, from arable land to urban gardens. Gorse is a favourite food plant, alongside dandelions, bluebells, dead-nettles, comfrey, selfheal and foxgloves.
How to identify These beautiful little bees are the only British species with all-brown colouring and no white tail. They range from ginger to a pale, sandy brown, depending on how sun-bleached they are.

Brown carder bees

When you'll see them out foraging March–November
Where Brown carder bumbles love clusters of flowers and wild vegetation like foxgloves, tussocks of dead nettles and legume flowers such as runner beans.

Tree bumble bee

How to identify They have varying shades of brown or ginger with bare rear legs that are shiny. Common carder bees have black hairs on their abdomen.

Tree bumble bee

When you'll see them out foraging March–July
Where As the name suggests, these bumbles love nesting in trees as well as bird boxes and nooks in buildings everywhere from woodland to urban gardens. Favourite floral foraging includes rhododendrons, brambles and comfrey.
How to identify Easily identified by their ginger thorax, black abdomen and white tail, tree bumble bees are now one of our most common species.

Shrill carder bee

When you'll see them out foraging May–September
Where One of Britain's rarest bumble bees, these peachy creatures are specific to grasslands in southern England and Wales that are rich in legume flowers such as vetches, trefoils and clovers.
How to identify The body of the shrill carder bee is pale yellow with grey bands. They have an adorable tail, which is peachy in colour and form – like the little furry point at the base of the summer stone fruit.

Solitary bees

Most of the bees in Britain are solitary bees. Of the estimated 270 different British species, solitary bees make up more than 90 per cent. Within this category are mining bees, leaf-cutter and mason bees. As the name implies, their distinguishing characteristic is that they don't live in colonies like honey and bumble bees.

Each female solitary bee builds her own nest. Depending on the species, these can be in the ground or in cavities high off the ground. The nest inside houses a series of egg cells, each protected by a wall of material and provisioned with nectar and pollen. Male eggs are laid towards the front of the entrance, so that they emerge before the females and are ready to mate in the following year.

Hairy-footed flower bee

When you'll see them out foraging March–June
Where One of the first bees to emerge in spring, darting rapidly between early blossom and flowers, they have a particular fondness for lungwort, dead nettles and wallflowers.
How to identify Small, bumble-bee lookalikes, the females are black with yellow legs, while the males are brown with pale faces and a plume of hair on their middle legs.

Hairy-footed flower bee

Green-eyed flower bee

When you'll see them out foraging April–July
Where Like the larger hairy-footed flower bee, this small flower bee darts quickly between a wide variety of flowers, emitting a high-pitched buzz. Look for them on lavender, catmint, hardy geraniums and flowering herbs.
How to identify Both males and females have distinctive green eyes, which make them easy to recognise. They are around 8–9mm in length.

Long-horned bee

When you'll see them out foraging May–August
Where This declining solitary bee is reliant on open habitats rich in legume flowers such as vetches and trefoils with nearby earth banks or cliffs for nesting.
How to identify These bees have grey-brown hair. The males have unmistakably large, oversized antennae. The females have a white tail.

Common mourning bee

When you'll see them out foraging March–June
Where These bees are parasites, laying their eggs in the nests of hairy-footed flower bees where they steal food stores.
How to identify Also known as the cuckoo bee, this species typically has a grey collar, white spots along the abdomen and a pointed rear. Sometimes they are fully black.

Wool carder bee

When you'll see them out foraging May–July
Where Females gather balls of plant hairs to craft their nests, which are typically built above ground, or they can be found nesting in man-made bee hotels. Males guard hairy plants such as lamb's-ear.
How to identify The wool carder bee has a black body with a stunning yellow pattern of half-moons along their sides and bottom. Males have three prongs on their rear.

Leaf-cutter bee

When you'll see them out foraging May–August
Where Leaf-cutters love roses, so if on your rose bushes you spy neat crescents cut from the edges of leaves or petals these bees are likely culprits. The female workers use the leaves to line their nests, which are typically found above ground or within man-made bee hotels.
How to identify This bee has a broad head and body. The female leaf-cutter bee has a brush of hairs under the abdomen.

Small scissor bee

When you'll see them out foraging June–August
Where The females adore diving into purple bellflowers (campanulas) for nectar before taking it to nests inside woodworm holes in dead wood. Male scissor bees can be found sleeping inside summer flowers. They also like elevated bee hotels.
How to identify The smallest of Britain's bees, at just 6–7mm, these little creatures are black and shiny with a slender body and a large head. The females collect pollen under their abdomen.

Sweat bee

When you'll see them out foraging April–October
Where Most species of sweat bee are solitary, preferring to nest on their own, typically in the soil, though they often nest close to others so they can share resources. Some species prefer to nest in wood rather than the ground. Sweat bees are so-called because they are attracted to human perspiration but won't sting unless provoked.
How to identify Among the smallest of Britain's bees, ranging from 4–10mm, they are mostly black, though some species have stripes, ranging in colour from yellow, though green to red. The females collect pollen under their abdomen.

Yellow-faced (or masked) bee

When you'll see them out foraging May–September
Where In gardens, hedges and forest clearings, feeding on a wide range of flowers.
How to identify These small black bees, around 4–7mm, have distinctive markings. The males have a pale yellow face mask, while females have just a small patch of yellow near each eye.

Gooden's nomad bee

When you'll see them out foraging April–June
Where Around the nests of other bees, typically mining bees, where they lay eggs and steal nectar and pollen. They don't collect pollen themselves.
How to identify Medium-large bees, with black and yellow stripes, often with the uppermost yellow stripe divided in two.

Mason bees

If you've ever noticed clouds of bees buzzing about in front of brick walls, they are likely mason bees. A solitary species that nests in cavities in wood, hollow stems and walls, mason bees look a little similar to some mining bee species, but you can tell them apart by their boxy heads and large powerful jaws. The red mason bee is the most common species.

Gooden's nomad bee

Red mason bee

Red mason bee

When you'll see them out foraging March–July

Where Also known as the two-coloured mason bee, red mason bees are typically found in built-up environments with plenty of gardens, churchyards and urban green space, and they are the bee most likely to be tucked up in your bee hotel. Their food plants include fruit trees, sallow and oilseed rape. Geographically, you'll see them mostly in lowland England and Wales. They're uncommon in Scotland and Ireland.

How to identify The red mason bee takes its name from its distinctive covering of ginger hair. Look out for a black head, brown thorax and orange abdomen, and in females, a lot of fluff and pollen under the abdomen.

Orange-vented mason bee

When you'll see them out foraging May–August

Where These bees love foraging plentiful patches of thistles and knapweeds and are keen to exploit man-made bee hotels. Geographically, they are common across southern England and the Midlands.

How to identify Males are tiny, about 6mm long; females are longer, about 10mm, but narrower. Both have bright ginger hairs on their belly.

Mining bees

If you've spotted a hole in your lawn surrounded by a volcano of excavated earth, it is most likely the work of a mining bee, a solitary species which nests in the ground.

Tawny mining bee

When you'll see them out foraging March–June

Where This bee makes volcano-like mounds of soil at its nest entrance in lawns and mown banks. Foraging females feast on shrubs ranging from willow, hawthorn and blackthorn to fruit trees and maples. They also love gorging on dandelions.

How to identify There is no mistaking the tawny mining bee – a honeybee-sized ginger species, which has a thick, bright orange coat with black stripes, a black face and black legs that contrast with the buttercup yellow of their pollen collection.

Ashy mining bee

When you'll see them out foraging March–June

Where Clusters of ashy mining bee can be found along footpaths and short turf on heathland, moorland edges, open woodland, coastal grassland, cliffs and quarries. They are important pollinators for oilseed rape.

How to identify This mining bee is primarily black with two light grey bands

Ashy mining bee

across the thorax. The glossy black abdomen can appear blue-black in the light. The female is larger with more distinctive markings than the male.

Orange-tailed mining bee

When you'll see them out foraging March–July
Where Widely seen in many habitats, including urban areas, these mining bees nest on grassy slopes and forage mainly from blossoming shrubs.
How to identify Also known as the early mining bee, this is one of the first bees to appear at the start of spring. The thorax is rusty orange and lightly haired, while the abdomen is black and fairly smooth, with a tuft of orange hairs on the tip of the tail. It also has unique yellow rear legs.

Ivy mining bee

When you'll see them out foraging September–November
Where A late-appearing solitary species that feeds solely on the pollen of flowering ivy. They nest in loose, light or sandy soil on south-facing banks and cliffs in spots with prolific ivy growth.
How to identify A honeybee dopplegänger, the ivy bee has an orangey-brown, hairy thorax with distinct black and yellow abdominal strips.

Ivy mining bee

Bee doppelgängers

While bees evolved from ancient wasps, their closest resemblance is the hoverfly. The bee fly, as the name suggests, is another lookalike and an imposter on several levels. The following trio are all brilliant pollinators, but they are carnivorous creatures, feasting on other insects – including bee larvae.

Hornet

When you'll see them out foraging April–September
Where Becoming more common recently, hornets typically build paper nests from chewed wood in rotten tree cavities, but they also look to nest in chimneys and similar voids. They're often seen as pests but they are great pollinators.
How to identify Our native hornet, the European hornet (*Vespa crabro*) has a cone-shaped yellow abdomen with black stripes and a reddish-brown thorax and legs. It is Britain's largest social wasp and is too big to be mistaken for a bee.

Hoverfly

When you'll see them out foraging March–November
Where Hoverflies are excellent pollinators and are often found collecting floral nectars from hedgerows, both in gardens and in woodlands.
How to identify There are two key identifying factors: hoverflies are quite tiny, and much smaller than bees; more telling, perhaps is that they do, indeed, hover, rapidly beating their wings as they linger in one place. If you look closely they have rather large eyes, which sometimes meet in the middle, and short antennae with few segments. The 'footballer' hoverfly (*Helophilus pendulus*) is a common species – its distinctive yellow and black striped thorax give it its name.

Bee fly

When you'll see them out foraging March–August
Where In bees' nests – they are parasites! Female bee flies fling their eggs into solitary bee burrows where the larvae eat the pollen stores.
How to identify Bee flies have a long proboscis and short antennae. Typically, they hover around flowers like dead nettles and spring primrose, and can also be spotted being very busy in the dusty earth. Unlike bees, they have a single pair of wings, rather than two sets. Otherwise, it is easy to be fooled into thinking they are honeybees. The dark-edged bee fly is the most common species in the UK.

European hornet

'The footballer' hoverfly

Dark-edged bee fly

Making Your Space More Bee Friendly

As cities grow and agriculture becomes more intensive, bees and other insects have lost much of their natural habitat. An estimated 98 per cent of wildflower meadows have disappeared from England and Wales since the 1930s, profoundly impacting our wildlife, including bees.

Wildflower meadows are richly diverse habitats, with a huge variety of native flowers. A single healthy meadow can be home to over 100 species of wildflowers, which in turn support other meadow wildlife. In the summer, an acre of wildflower meadow can contain three million flowers, collectively producing around 1kg of nectar in a single day, which is enough food to support nearly 96,000 honeybees.

As wildflower meadows teeter on the edge of vanishing from our landscape, some species of bee have evolved to only eat a limited number of pollens, but this can result in a less nutritionally diverse diet for the bees. Also, when flower species are limited, it results in too much competition, i.e. not enough food for the wildlife who depend on it and, thus, species (including bees) reduce.

Hedgerows are also in a worrying state of decline, further reducing a varietal forage for bees. Since World War II, we've lost around 50 per cent of them. Hedgerows provide an abundance of food, from the little clusters of dazzling white hawthorn blossom and small axillary flowers of holly or yellow gorse (an important overwintering food for bees that venture out on a sunny day) to the fuzzy pussy willow, which arrives in pollen-scarce months, and the starry splay of alder buckthorn blossom in summer. Hedgerows are also important for human health, as street-edged hedges are a natural barrier to harmful pollution. On farms (and in gardens), maintaining undisturbed hedgerows can encourage pollinators to help increase pollination and thus productivity for other plants, such as squashes, blueberries and tomatoes.

And with gardens alone, there's so much you can do to foster healthy habitats for bees, the most important being to plant a wide range of bee-loving flowers (see pp.52–5). There are about 22 million gardens in the UK, covering 400,000 hectares. If we work together to sow lots of insect-friendly seeds, lay off the chemicals, provide water and create some wild habitats, we could make an enormous difference to insect abundance. The single most constructive thing that anyone can do for bees is to create better, safer habitats.

This includes caring for your soil, too, as wildlife doesn't just exist above ground. Looking after your soil but practising no-dig and feeding it with organic compost will foster the life beneath our feet (one gram of soil can harbour up to ten billion micro-organisms) from which springs healthy plants for bees to feast on.

Habitats for different bee species

Knowing what different bees seek for shelter and how they live once inside their abode will help you be of assistance. Increasing the diversity of their forage naturally opens up housing opportunities for bees as some species, male bumble bees, for instance, lead a vagabond life as they don't have a colony or nest to return to. Often, they use the cup of a flower as a cradle at night.

Should you want to house bees in more man-made structures, like modern honeybee hives or shop-bought bee hotels, it's important to conscientiously look after your apian guests, assuming the role as a guardian of their welfare by ensuring any man-made home is free of diseases, fungi, parasites and predators (such as mites, wasps, flies, spiders, snakes and mice). Without proper care, bee hotels can unintentionally be unhealthy habitats.

A brilliant way to really care for bees is to provide naturally occurring nest sites (logs, clusters of sticks and tufts of grass) and bees quite possibly will come. And if they do, ensure they're not exposed to chemicals, try to protect them from predators and ensure they have access to food (plenty of year-round flowers) and shallow sources of water (too deep and they can drown).

Honeybees

In the wild, honeybees make their nests in large cavities, usually hollow tree trunks or even in caves but they're also known to seek out more unconventional sites such as chimneys, walls and roofs of buildings, or even unlikely places such as hollow statues, compost bins and graveyard crypts.

Once nestled inside their chosen cavity, the female worker honeybees start building hexagonal comb so the queen can commence laying eggs to keep the colony in perpetuity.

In the height of summer, when the hive is ripe with reproduction, a colony of bees can swell to 50,000 bees. It's a city of sorts with a fascinating hierarchy consisting of three different tiers: the queen (of which there can only be one), drones (the males in the hive) and worker bees (the largest population in the hive, consisting of all females with many different jobs within the colony).

These different demarcations determine life-span. The queen bee reigns the longest – for up to 4 years, and sometimes longer. Drones, which are always male, have a life-span of about 55 days and have a sweet life. They spend their days lapping up the honey brought in by the worker bees (the females) and meet their end immediately after successfully mating with a queen. Basically, their job is to eat and have sex. Worker bees, it seems, get the short end of the stick. They're put to work the moment they hatch and assume various roles within the hive in their fleeting existence. They typically live just 35 days, toiling away.

'Natural' colony of honeybees living in a hole in an ash tree

The colony shrinks down in winter to as few as 5,000 bees. The drones are kicked out of the hive as they provide no service. The winter posse is just workers surrounding their queen, huddling together to stay warm until the temperatures begin to creep back up to 10°C, when they start to venture out to gather spring nectar, and thus the cycle begins again.

Bumble bees

Fat and fuzzy bumbles have all manner of housing preferences and they're pretty creative when it comes to making the most of wild spaces, setting up home in unexpected spaces like nooks behind vents or compost heaps, while some species prefer high up spaces like people's lofts or bird boxes.

It's up to the bumble bee queen to find the perfect nesting spot to lay her eggs. The queen lives for an annual cycle and in that time, she builds up a colony ranging from 30 to as many as 400, depending on the species.

As the summer peaks, so does the colony size. The queen will start laying unfertilised eggs, which develop into male drones, while the workers (young female bumble bees) feed the fertilised eggs laid by the queen with royal jelly so they develop into new queens suitable to establish a new reign.

These new bumble bee queens (called gynes) will mate in the autumn and then hibernate alone before starting their own individual colonies the following spring.

Examples of low-lying bumble bee nests

- Underground nooks like abandoned mouse holes
- Crevices caused by tree roots
- Holes or cracks in rocks and walls
- Tussocks of grass
- Compost heaps

Examples of aerial bumble bee nests

- Crevices in trees
- Holes in walls
- Cavities behind vents
- Bird houses
- Loft spaces

Bumble bee species that use pre-existing holes for nests may make use of pieces of dried grass, dried leaves and perhaps mouse hairs or dried matter they find inside the crevice.

Solitary bees

Around 90 per cent of the world's bees are solitary species and the solo-dwelling mining bee is the largest genus of bee in the United Kingdom. Rather than seeking shelter in treetops or living as a colony in a hive, solitary bees (as the name implies) prefer a solo abode, typically low to the ground: a burrow dug into the soil or in a clay tunnel, or a nest in the hollow of a plant stem, or hole in wood or brick. They also don't make honey.

The cycle starts from the first mild days of spring when male solitary bees become active, feeding on early nectar and buzzing restlessly around potential nesting sites, waiting for the females to arrive. A male solitary bee will mate with several female bees, until its supply of sperm is exhausted.

When the female bee emerges from her individual nest, created by her mother, she'll find a male to mate with and then seek a nesting site of her own. Once nesting sites are spotted, she'll gather pollen in bundles stuck together with nectar, as honeybees do. Each nectar-glued bundle of pollen will serve as a nutritious food parcel for each of the eggs – she'll lay around 30 eggs. Sometimes a nesting site will house a few eggs, but each will be partitioned with a mixture of soil, nectar and saliva to give each egg a nesting space of its own. The nesting site is then closed with mud, leaves or fine hairs.

The mother bee has served her purpose and her eggs will hatch into larvae after a few weeks and feed on the provided larder. By the autumn they are fully grown and remain dormant until the following spring. The larvae pupate in their cells and then emerge as fully adult insects in May or June when they leave the nest to continue the annual solitary bee cycle.

Pesticides: a major threat to bees

Of all the threats to bees, pesticides are arguably the biggest. The main offender in recent times is neonicotinoids (aka 'neonics'). According to academic and author, Professor Dave Goulson, one teaspoon of neonics is enough to kill 1.25 billion honeybees, equivalent to four lorry loads.

Neonics are a class of insecticides, similar to nicotine in the way they act on the central nervous system. They can disrupt a bee's ability to fly and navigate, which impacts its ability to gather nectar and pollen to feed budding bees in the hive. Neonics were developed in the 1980s by scientists at Shell, the British oil and gas company, and Bayer, the German pharmaceutical and biotechnology company.

Neonicotinoids are often applied to seeds before planting as a treatment against herbivorous insects. They are water-soluble, so when the seed sprouts and grows, the developing plant absorbs the pesticide into its tissues as it takes in water.

Neonics can also be applied to the soil directly and once the soil drinks it in, the toxins become present throughout the plant, including in its leaves, flowers, nectar and pollen. The latter is why their use has been linked to the rapid decline in honeybees, bumble bees and other insects, as well as insect-eating birds.

With growing evidence of their harm on wildlife (bees, in particular), some neonicotinoids were banned for outdoor agricultural use in the UK and the EU in 2018. However, since Brexit, exceptions have been made for their use in the UK, with the British government allowing for the use of thiamethoxam – a type of neonicotinoid – on sugar beet in England. The move to support the use of environmentally damaging chemicals to increase the production of sugar, of all things, is upsetting to say the least.

But neonicotinoids aren't the only pesticide problem for pollinators. Bees are exposed to a whole range of different pesticides, including insecticides, fungicides and herbicides. A growing body of evidence is showing that pesticides can become even more harmful when used in combination. The Cocktail Effect report, released by the Soil Association and Pesticide Action Network UK in 2019, demonstrated that UK soil and water contained mixtures of ten different chemicals, with the potential to affect wildlife, like birds and bees. Detailing the cocktail effect phenomenon, the report describes how combinations of different pesticides synergise, interacting to worsen their impact on bees. For instance, fungicides don't always harm bees on their own, but they can make certain insecticides up to 1,100 times more potent.

Humans have used pesticides to protect their crops for over 4,500 years. The first known pesticide was elemental sulfur dusting used in Sumer around 2500 BC in ancient Mesopotamia. But it was World War I that hastened the industrialisation of pest control and post-World War II (1940–50s) is deemed the start of the pesticide era. Pesticide use has increased 50-fold since 1950. In 2021, global pesticide consumption was 4.2 million metric tonnes.

While many insects can develop a resistance to pesticides, bees cannot, because of the special relationship between bees and flowers. When locusts, beetles, aphids, Lygus bugs and other pestiferous creatures attack leaves, stems, seeds and root, they do so because their existence relies on detoxifying complex compounds. For millions of years, non-pollen- and nectar-seeking insects struggle (but manage) to overcome the constantly evolving chemical defences of the plants they feed on.

Bees are different. Their role as pollinators creates a need for plants to attract them, not repel them, and the sweet nectar and protein-rich pollen plants have evolved to produce flowers to attract bees. But, unfortunately, bees are unable to deal with the plant's defensive chemicals. This means bees are lured to food that can potentially poison them. Once exposed to pesticides, bees have increased susceptibility to parasites, poor foraging and reproduction.

While this pesticide information might make for a grim read, there's a very simple solution that's good for bees, as well as all forms of wildlife and us, too – we are all part of the food chain, thus linked, after all. One of the simplest things you can do to help combat the increase of pesticide use is to support organic farming. This doesn't only apply to food crops, take clothing: non-organic cotton is considered the world's dirtiest crop due to its heavy use of chemicals, using 16 per cent of the world's pesticides annually as crops are sprayed up to 20 times in one growing season.

Farmers on organic farms are permitted to use just 20 pesticides, compared to around 400 in non-organic farming. These 20 pesticides are derived from natural ingredients, like citronella and clove oil, and they are only permitted under very restricted circumstances. One of the key focuses for organic farming is to create diversity, which helps create a healthier soil, which in turn promotes stronger, more resilient plants that are less susceptible to pests. You can look at pests like weeds – they're just insects in the wrong place – and if you provide them a home they won't attack your crop. As a result, there are around 75 per cent more wild bees on organic farms and overall, plant, insect and bird life is 50 per cent more abundant.

Planting for bees

Imagine if your diet contained nothing but almonds, just apples or only squash. This is a picture painted by author Thor Hanson, who wrote about the need to support all our bee species in his brilliant book, *Buzz: The Nature and Necessity of Bees*, alluding to the monocultures some bees (a practice common in America) experience when they're shipped to assist with crop pollination. Like us, he outlines, bees want and need a diverse diet to get a range of nutrients to keep their immune systems strong and to fight the increasing onslaught of environmental toxins.

When bees are presented with the narrow forage of a monoculture, they will go looking for other sources of protein and micronutrients because they know (almost better than us humans) what they need to stay healthy. 'The mess that we're in now is a landscape basically devoid of flowers,' according to Dave Goulson, Professor of Biology at Sussex University, who is specialising in bee ecology. 'It's vital that we increase the extent and richness of plant communities,' especially if you want to have your own bees. 'If all you're doing is adding more honeybees, you're very likely to be doing harm.'

There are huge benefits in this diversity for us all. We, too, need to widen our own forage (and eat a much more diverse spectrum of plants). Our health is failing as a result. But when it comes to bees, studies show that the presence of more wild species leads to an increase in harvest yields.

Verbena bonariensis

In short, the best thing you can do to support our pollinators is plant wildflower-rich habitats. If you don't have your own garden, get involved with community gardens, local parks or groups creating bee-focused habitats.

Why bees like certain flowers

Plants with pincushion-like flowers such as Knautia, Astrantia and Echinacea, which have numerous small flowers bundled together on one head, are excellent forage for honeybees as they have a shorter proboscis, which is the 'straw' used by the bee to drink the nectar. Plants with lots of individual flowers held on spikes such as lavender, Liatris, Veronicastrum, also provide easy forage for honeybees. But that said, there are some exceptions. Salvia, for instance, is so attractive to honeybees and bumble bees that they sometimes create a hole in the base of the flowers to access the nectar from the slender, long-necked flowers.

Avoid using chemicals

Try to garden as organically as possible, avoiding the use of chemicals and opting for natural, organic alternatives instead. When buying seed, check that the plants and bulbs have not been pre-treated with a cocktail of bee-harming poisonous chemicals. Labels on plants don't always give much away so it's best to buy from organic suppliers, as organic seeds and bulbs will be chemical-free.

A great way to get your hands on bee-friendly plants is swapping from trusted sources, such as community gardens or neighbours practising organic approaches. Sowing and/or saving your own seeds is another great idea. Feed the soil, compost, mulch and allow pests to build up just enough to encourage the beneficial insects and predators. If your garden is alive, it will sustain a substantial family of bees and other insects, invertebrates, small mammals and birds.

Think in colour

Bees see differently to us. They can't detect red, but blue and violet lure bees and they see ultraviolet, which is hidden to us. Ultraviolet often signals where the nectar is to be found. Some plants help further, with the colour changing once the flowers are pollinated. Pulmonaria, for example, changes from blue to pink. And forget-me-nots lose their yellow ring once pollinated; it fades to a creamy brown, signalling to the bees that there is no longer any nectar. There are some interesting studies using UV photography that highlight patterns in flowers (like landing strips!) that are not visible to us – but seen by bees.

A simple guide to planting for bees is to think purple. If a bee were to name a favourite colour, the regal hue would be their named choice. Linger in a herb garden and you'll see their preference in action – the chive blossom, verbena blooms and violet-shaded thyme flowers will be the ones buzzing with a hive of activity.

Bees are herbalists

Knowing what's healthy for them, bees will fly up to 3 miles away from their hive in search of the right pollen to enrich their vitality. When it comes to honeybees, all that goodness is imbued in their honey, too.

If you dive into the River Cottage kitchen garden to gather herbs on a sunny day, you're serenaded with the gentle hum from several bee species gathering nectar from these medicinal plants. They know what's good for them, and happily what's good for them is also good for us. If you take a gut health test and discover you have an imbalance of good versus bad bacteria, most nutritionists prescribe a thyme tincture to help rid of the bad bugs. The same volatile oils (thymol) are also used in varroa mite treatments. When bees gather nectar from the thyme flowers, they're instinctively striving to stave off varroa mites.

Bees have also been observed foraging the mycelium of fungal species, such as wine cap. Experts are now taking their lead, looking at medicinal mushrooms such as reishi to help treat viruses affecting colonies.

While humans also have intuitive instincts towards healing foods, our attention is so diverted by marketing, processed foods and lack of connection to nature that we literally can't see the wood for the trees. Observing what bees dive into can help reacquaint us.

Herbs to plant for bees

A carefully planned herb garden, pot or potager – where herbs and edibles are mixed in with decorative, ornamental planting – is beautiful and can provide aromatherapy, while also offering delicious ingredients for us and the bees.

It's traditional practice to cut your herbs back before they fully go into flower, as the herbaceous leaves can go woody and sometimes a touch bitter. But, it's worth growing extra plants so you can leave some to flower, while cutting the others back for continual leaf harvesting. Having herbal flowers in the garden will equate to their flavours and medicinal properties in the honey (for the bees, and us if you're able to harvest some).

Angelica

This can grow to 2 metres, so plant it where you need a bit of height or drama in the garden. It has purplish stems and small white flowers with a honey-like aroma.

Months in flower June–July
Medicinal purpose for bees Contains anti-fungal chemicals
Attracts Hoverflies, solitary bees and sweat bees

Basil

Any variety of basil that blooms is a good source of pollen and nectar for the bees. Thai basil, with its purple-tinged stems and leaves, looks great in the garden. Don't pinch the flowers to encourage bushy growth, but rather focus on producing enough for yourself and the bees.

Months in flower July–September
Medicinal purpose for bees Helps fight parasites
Attracts Honeybees (nectar and pollen)

Bergamot

Bergamot is also known as bee balm, because the plant (leaves and flowers) is used to treat bee stings. Bees, especially bumble bees, are completely obsessed with the magnificent pink flowers, which are large, complex, rich in nectar and held on strong stems.

Months in flower July–August
Medicinal purpose for bees Helps fight infections and supports bees' immune systems
Attracts Bumble bees

Borage

This herb has dainty, star-shaped blue flowers and a long flowering period, often filling a usual gap in the supply of nectar for bees in mid-summer, which beekeepers refer to as the 'June gap'.

Months in flower June–August
Medicinal purpose for bees A rich source of calcium, iron, potassium, zinc, vitamin C and beta carotene
Attracts Honeybees, for the nectar; bumble bees, including the common carder bee, early bumble bee, buff-tailed bumble bee and white-tailed bumble bee for the pollen

Chives

An easy-to-grow perennial, which means it comes back year after year, chives offer a long flowering period.

Months in flower May–August
Medicinal purpose for bees Helps prevent and treat bacterial and fungal infections
Attracts Bumble bees, honeybees, mason bees and leaf-cutter bees foraging for nectar

Honeybee about to collect nectar from borage

Comfrey

This is a hardy perennial (so comes back year after year), with a long flowering period. The bees love it, and it can be chopped down regularly as a feed for your compost. Biologist and bumble bee expert, Dave Goulson, deems comfrey one of the 'very best plants for bees'.

Months in flower May–August
Medicinal purpose for bees Rich in lots of nutrients, especially protein, nitrogen, phosphorus and potassium (making it a brilliant food for your garden, too)
Attracts Long- and short-tongued species, the latter often robbing from holes bitten in the tops of the flowers

Coriander

Letting your coriander plants flower in the garden not only provides bees with an excellent source of protein-rich pollen, it also gives you a host of dainty white flowers that will turn into seeds; these are delicious eaten fresh, offering a mildly spicy flavour and a citrusy zing.

Months in flower June–July

Medicinal purpose for bees Immune-boosting, calming and rich in nutrients (especially protein-powered pollen)
Attracts Honeybees and bumble bees

Dill

An attractive plant with soft feathery leaves and flower stems stretching to 90cm tall, producing delicate clusters of small yellow flowers which bees love.

Months in flower June–August
Medicinal purpose for bees Rich source of pollen, offering protein
Attracts Honeybees and hoverflies

Echinacea

One of the prettiest flowers in the summer garden, echinacea (aka 'coneflower') typically have long pink daisy-like petals that droop down like the ears of a basset hound from the open heart of the flower, which offers bees easy access to its impressive variety of immune-boosting compounds.

Months in flower June–September
Medicinal purpose for bees A rich source of caffeic acid, alkamides, phenolic acids, rosmarinic acid and polyacetylenes, which are support for both human and insect immune systems, helping them to fight off infection
Attracts Many different species of bees, including some sweat bees, bumble bees and honeybees

Fennel

Bronze fennel and other varieties that don't produce large bulbs are wonderfully ornamental, the wispy fronds, pollen-rich yellow blossom and fresh aniseedy seeds taste amazing and the herb is a complete honeybee magnet.

Months in flower June–August
Medicinal purpose for bees Immune-boosting
Attracts Honeybees, which are drawn to fennel in droves, but this frondy herb is also a host plant for swallowtail butterfly caterpillars

Hyssop

An evergreen herb in the mint family, hyssop is popular in herbal medicine due to its purported properties as an antiseptic, cough reliever and expectorant.

Months in flower July–September
Medicinal purpose for bees Kills infection
Attracts Honeybees, bumble bees, leaf-cutter bees, and yellow-faced bees

Honeybees attracted to echinacea

Lavender

If you want your garden buzzing with fluffy-bottomed bumble bees, fill your patch with lavender. Research from Sussex University shows bumble bees are ten times more likely to visit lavender over borage, whereas honeybees are drawn to the latter, starry-blue flowers.

Months in flower June–August
Medicinal purpose for bees The pollen is rich in amino acids
Attracts Bumble bees, which are especially fond of lavender because their longer tongues are better at extracting nectar from the narrow blooms, but honeybees love it, too

Lemon balm

Another word for lemon balm is Melissa, which is the Greek word for bee. Some say if you put a sprig of this herb in an empty hive it will soon be full of bees. We haven't tried it with lemon balm sprigs (yet) but a few drops of lemongrass oil certainly have worked to lure bees into our hives.

Months in flower June–August
Medicinal purpose for bees Lemon balm produces oils which mimic the Nasonov pheromone produced by honeybees, and in doing so calms the bees and prevents swarming
Attracts Bumble bees, who commonly consume its nectar, which is less accessible to shorter-tongued honeybees

Oregano

Due to its bloom in the height of summer and abundance of nectar, oregano draws in a great number of bees and range of species. A study of the attraction of bees to several shrubs, which took place in Jordan, showed that over the course of 42 days over 21 species of bees visited the Syrian oregano.

Months in flower June–July
Medicinal purpose The vast supply of nectar oregano offers to bees is also seen to have antibacterial qualities.
Attracts A broad spectrum of species, but this herb is particularly enticing to short-tongued honeybees and bumble bees

Mint

There are numerous species of this bee-friendly herb that have clusters of short flowers, which enable bees to consume pollen and nectar quickly without having to travel far between flowers.

Month in flower July
Medicinal purpose for bees Studies have shown that the nectar and pollen from spearmint can help clean out mites that infect the hives
Attracts Most bee species and many other pollinators, owing to the ideal structure of its flowers

Rosemary

The pale purple flowers of rosemary offer a fantastic feast to bees, as they provide both pollen and nectar. Rosemary flowers from the beginning of spring, so it offers a source of early food for establishing colonies.

Months in flower April–June
Medicinal purpose for bees Rosemary is believed to combat mites and particularly varroa mite, which is a common enemy among honeybee colonies
Attracts Honeybees and bumble bees

Sage

The aromatic plant of sage produces petite purple flowers, which provide bees with sources of nectar and pollen. Russian sage is notably rich in both. Honeybees capitalise on this combination of foods by stuffing the pollen in the baskets located on their legs and absorbing the nectar with their tongues.

Months in flower June–September
Medicinal purpose for bees Sage provides bees with minerals such as cobalt, iron and magnesium. These elements then translate into the honey they produce
Attracts Leaf-cutter bees and bumble bees, as these species have tongues long enough to reach the flower's treasures

Thyme

Bees love all varieties of thyme as the flowers are easy to access and the herb has volatile oils, including thymol which is used in varroa mite treatments. They have a particular penchant for lemon thyme because they adore the scent of lemony fragrances.

Months in flower April–July
Medicinal purpose for bees When bees gather nectar from thyme flowers, they're instinctively ingesting a protective oil to help keep varroa mites down
Attracts Honeybees, bumble bees and leaf-cutter bees, in particular, but the small flowers are accessible to bees of all tongue sizes.

Honeybee collecting nectar from marigolds

Bee-friendly plants for every month

Bees need flowers for food and energy, and flowers need bees and other pollinators to reproduce; it's a beautiful symbiotic relationship set in motion a hundred-million years ago and it has been evolving ever since. Flowers spring from trees, bushes and shrubs lining hedgerows or on their own. They emerge as single stems and from several edible plants on farms and in gardens. Plants often considered weeds are good sources of food for bees and other pollinators, too. Below is a guide to plants to cultivate to help ensure bees have abundant food resources throughout the year.

January

In this mid-winter month the number of bees in the sky is sparse, as the masses cluster together in their hives, shivering their flight muscles with the aim of generating heat. This will allow them to survive through the coldest months. Though, there is one exception in the UK: buff-tailed bumble bees, who continue to harvest throughout the winter.

Bee-friendly plants in flower

- Winter heath heather
- Winter aconite
- Clematis cirrhosa
- Viburnum tinus
- Hazel
- Snowdrops

February

In this month, the buff-tailed bumble bees are joined by the hairy-footed flower bees, with the males emerging first and the females a few weeks later.

Bee-friendly plants in flower

- Goat willow
- Spring flowering crocus
- Snowdrops
- Erica x darleyensis
- Cherry plum

March

The arrival of spring welcomes bees such as early bumble bees, red mason bees and the orange-tailed mining bee, emerging after a long winter. Having blooming plants in this month is vital to giving the bees their first taste of pollen for the year.

Bee-friendly plants in flower

- Primrose
- Flowering currant
- Lungwort
- Helleborus x hybridus
- Dandelion

April

As the year progresses into April, more bees start to surface with the tawny mining bee, Gooden's nomad bee and ashy mining bee, as well as a further influx of bumble bees, all starting to feast upon flowers.

Bee-friendly plants in flower

- Apple
- Fiddleneck
- Granny's bonnet
- Wood spurge
- Berberis
- Cherry

May

May is one of the earliest months when the bumble bees you see transition from queen to worker bees, which have the primary job of gathering nectar and pollen. In this month you will also start to notice worker honeybees collecting from the same plants as bumble bees.

Bee-friendly plants in flower

- Elder
- Nettle-leaved bell flower
- Ceanothus
- Comfrey
- Brambles
- Green alkanet

June

The arrival of summer sees some of the most striking bees emerge, such as the green-eyed flower bee and the distinctive red-tailed bumble bee. Look out for these species – identifiable by their obvious features.

Bee-friendly plants in flower

- Allium
- Lupines
- Foxgloves
- Cardoon
- Common poppy
- Evergreen clematis

July

At the height of summer most bees are out in action. Among them, some of our smallest bees emerge this month, including the small scissor bee, which at 6–7mm is the smallest bee species in the UK.

Bee-friendly plants in flower

- Roses
- Lavender
- Angelica
- Marjoram
- Sea holly
- Marigold

August

Bees now focus on collecting nectar as they start to prepare for autumn and winter. By growing plants which bloom in August you're helping bees stock up for winter.

Bee-friendly plants in flower

- Teasel
- Field scabious
- Globe thistle
- Catmint
- Fuchsia
- Wild carrot

September

Late-blooming plants provide diversity to the bees' winter store as many crops thrive in this last blast of sun.

Bee-friendly plants in flower

- Salvia 'Amistad'
- Buddleia
- Verbena bonariensis
- Single-flowered dahlias

October

As we enter peak harvest time, the bees are starting to prepare for winter inside the hive, which means you tend to see fewer bees flying about.

Bee-friendly plants in flower

- Ivy (*Hedera helix*)
- Perennial wallflower
- Abelia x grandiflora
- Strawberry tree (Arbutus unedo)

November

The annual cycle has now gone full circle as bees in the sky are once again sparse. The only species around are honey, bumble, common carder and ivy mining bees.

Bee-friendly plants in flower

- Winter-flowering honeysuckle
- Sweet box
- Fatsia japonica
- Elaegnus submacrophylla
- Autumn-flowering crocus
- Ivy (*Hedera helix*)

December

At the height of winter, buff-tailed bumble bee and honeybee workers are the only bees who will occasionally brave the cold, while other bees shelter to survive.

Bee-friendly plants in flower

- Mahonia
- Winter-flowering honeysuckle

Understanding Honeybees

The different types of bees within each species: queens, drones (male) and workers (female), each have their own variations in shape and size. Queen honeybees, for instance, are longer than worker honeybees, while drones are plumper than worker honeybees (almost double the size) and have bigger eyes. Not all bees can sting: male bees, for instance, don't have stingers and cannot do you any harm. In this chapter we look at the roles off the different bees within a hive and behaviours that are critical for the colony to thrive.

Becoming familiar with the physical attributes of honeybees, and their patterns of behaviour, will enable you to carry out detailed inspections if you decide to nurture a colony of bees in your own garden. It will also tune you in to the fine details of this remarkable species and help you differentiate between individual types of honeybees, bumble bees and bee-like flies (such as the delicate, artfully striped and wide-eyed hoverflies).

Honeybee anatomy

In common with most other insects, bees have a hard outer shell called an exoskeleton and three main body parts: the head, thorax and abdomen. Honeybees also have two antennae extending from their head, three pairs of legs for walking and two pairs of wings for flying.

Head

This houses the brain and sensory organs – the eyes, mouth and antennae. It is also, of course, where food is ingested.

Antenna A movable segmented feeler that detects airborne scents as well as temperature, humidity and wind speed; it has the capacity to feel as well as taste
Ventral nerve cord The large bundle of nerves from the brain that sends signals to the rest of the bee's body
Compound eye A type of insect eye that is made up of many light detectors called ommatidia that give a wide field of view and provide motion detection
Ocellus A type of insect eye used to detect the intensity of light for navigation
Food canal The opening through which the bee takes in food (akin to the human mouth). Bees' food is almost always liquid in the form of nectar or honey
Labrum The mouthpart that forms the top of the feeding tube and can help handle food
Mandibles Pair of jaws used to chew pollen and work wax for comb building. They also help with anything that the bee needs to manipulate
Maxilla Mouthpart beneath the mandible that can handle food items

The anatomy of a honeybee

Head
Thorax
Abdomen
Hind wing
Fore wing
Ocellus
Antenna
Sting
Corbicula (pollen basket)
Compound eye
Hind leg
Proboscis
Fore leg
Middle leg

Honeybee castes

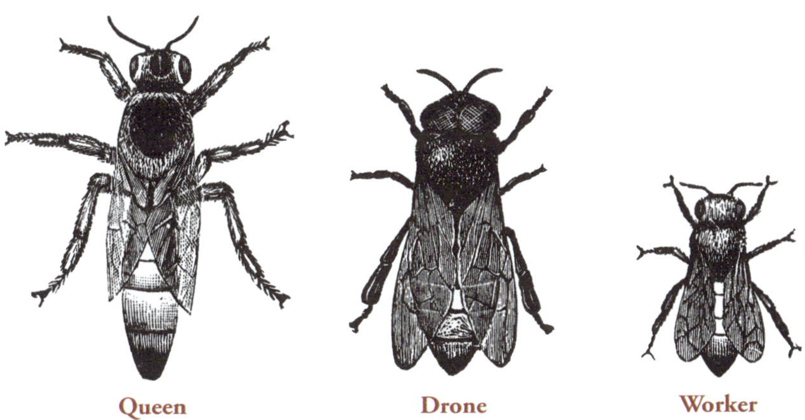

Queen Drone Worker

Labium A tongue-like appendage used to help drink up nectar that fits inside the maxilla like a straw. Bees can taste with this organ
Pharynx Muscles used to move the labium and suck up nectar from flowers
Labial palp Mouthpart used to feel and taste during feeding
Glossa A bee's hairy tongue that can stick to nectar to pull it in towards the mouth
Proboscis Straw-like mouthparts of a bee used to drink fluids
Oesophagus Part of the bee digestive system that begins below the mouth and extends through the thorax, connecting to the honey stomach in the abdomen.

Thorax

The midsection, where the three pairs of legs and two pairs of wings are attached.

Forewings The pair of wings closest to the head
Hind wings The pair of wings slightly further from the head
Forelegs The pair of legs closest to the head
Antennae cleaners Notches filled with stiff hairs that help bees clean their antennae. There is one on each foreleg
Middle legs The pair of legs located between the foreleg and the hind leg
Hind legs The pair of legs located furthest from the head. In workers, these legs have a unique set of tools used to collect and carry pollen. Located on the metatarsus (the fifth segment of the leg), the pollen brush has fine hairs, which are used to collect the pollen from flowers before moving it upwards to the tibia – the fourth segment (corbicula) where the pollen basket is located
Pollen basket Located on the tibia of the hind leg, this is an arrangement of fine hairs used for carrying pollen to the hive. Only worker honeybees have these (other bees carry pollen on their bodies). Having gathered the pollen with the pollen brush and moistened it with saliva, they pack it into the pollen baskets. A forager laden with full pollen baskets is easy to spot if you're inspecting the hive entrance: the bees will have a yellowy-orange powder attached to the backs of their legs.

Abdomen

The hind part of the bee, which houses the bee's stomach, honey sac and, in female bees, the reproductive organs and stinger.

Stinger Not all bees have stingers but those that do are always female. The stinger is a sharp organ protruding from the end of the female bee's abdomen, attached to the venom sac. An important feature of the worker bee sting is a microscopic pattern of barbs along its length
Venom sac The sting itself is attached to a venom sac inside the bee which contracts and pumps venom into the intruder.

Honeybee castes and behaviours

In a honeybee colony, all the bees are assigned specific roles, which collectively enables the colony to thrive. This behaviour is crucial, as individuals cannot survive on their own. In this chapter we will look at the assignment of roles, and behaviours that are crucial for the colony's success.

In the world of honeybees, a caste refers to a distinct form/type of bee that has a particular set of roles in the hive. The three castes in a hive are the drones, workers and the queen. Drones are male bees, the queen and the workers are female, but only the queen is fertile. They are all vital to the success of colonies, but perform very different roles.

The three castes differ in appearance and behaviour. The workers are smaller, have a sting and are very numerous. The drones are longer, wider, stingless and have much larger eyes; their population is 10 per cent or less of that of the workers. The queen is the largest bee, with a long abdomen, and has a sting. There is only one queen in a hive.

Each caste has a different time period elapse from being laid by the queen as an egg, to emergence as an (almost) fully formed bee. These timings and crucial events are detailed on p.100.

The queen is a fertile female and carries sperm from drones collected on her mating flight. This sperm is stored for her lifetime – potentially up to 7 years. When she lays an egg in the hive it can be fertilised using the stored sperm, or not. If fertilised, the resultant bee will be a worker (female); if unfertilised the bee becomes a drone (male). Female bees therefore have genetic material from both male and female bees. Their cells are referred to as diploid and have 32 chromosomes: 16 from the queen, and 16 from a drone. These worker bees may not be full sisters due to the queen carrying sperm from many drones.

The drones, however, only have genetic material from a single parent – the queen. So, their cells are haploid and contain a single set of chromosomes. This results in every sperm from a drone being genetically identical, i.e. they are clones.

Worker bees' tasks in the hive

The vast majority of bees in a colony are workers, and they undertake almost all of the activities required to sustain the colony. The number of worker bees varies dramatically through the year, dropping from around 50,000 in peak season to 5,000 or fewer in the winter.

The worker starts as a fertilised egg, laid by the queen. After 3 days this egg hatches into a larva, which is fed royal jelly, honey and pollen by the other workers and rapidly grows to fill the width of the cell in the comb. On day 9 the cell is capped with wax, and the larva transitions to a fully formed bee over the following

12 days – emerging by eating its way out from the capped cell on that day 21. From this day onwards, the workers perform different tasks within the colony that change as they age/mature (this is known as temporal polyethism). The actual timings are quite flexible, and overlap, but approximate to the following pattern:

- 1–2 days: Cleaning duties and keeping the brood warm
- 3–5 days: Feeding the older larvae
- 6–11 days: Feeding the younger larvae
- 12–17 days: Producing wax, building comb, carrying food, removing dead bees
- 18–21 days: Protecting the hive entrance from intruders, i.e. guard duty
- 22 days until death: Collecting resources for the hive, i.e. foraging for pollen, nectar, propolis resin and water.

Let's review the workers' age-related roles in and around the hive in more detail:

Cleaning and polishing

This role focuses on preparing pre-used cells in the comb for older workers to store pollen and nectar, and for the queen to lay eggs.

Feeding the larvae

In the first week or so after emerging, the worker's hypopharyngeal gland has not fully matured to provide the nutrient-rich royal jelly secretions that are used to feed the youngest larvae. So, during this period, the worker feeds the older larvae with a mixture of pollen, nectar and royal jelly. (Royal jelly is a complex mix of water, proteins, sugars, fats, vitamins, salts and amino acids.)

Once its hypopharyngeal gland has fully developed (around day 6) the worker moves to peak production of royal jelly and feeds it to the youngest larvae in the hive. Worker, drone and queen larvae are all fed the same diet for their first 3 days. The young workers are kept extremely busy continuously feeding the larvae as they rapidly grow.

Producing wax

Once sufficiently mature, and with a diet of honey/nectar, the workers become capable of producing wax. This is excreted as flakes from four pairs of glands in the abdomen and is then chewed and used to build comb. Each worker only produces a minute amount of wax each day, but the large number of workers can build comb very quickly when hive temperatures and incoming nectar are optimised.

It's worth noting that 1kg of wax is composed of approximately 1 million wax flakes, which requires at least 6kg of honey to be consumed to produce them.

Eggs laid by the queen in a new section of comb

Young hatched larvae

The workers assigned to this task will mostly make cells of wax in the comb that are suitable for workers to grow in. But they also make enough larger cells for drones, and a few 'play cups' which are shaped like an inverted cup and can occasionally be used to produce a new queen.

Removing dead bees

The colony needs to be kept hygienic to reduce the risk of disease. As there are many thousands of bees in a colony, it is inevitable that some will die within the hive – these need to be removed to avoid contamination. It is common to see workers carrying dead bees (and larvae) out of the hive and dropping them outside.

Carrying food

As workers mature, further duties include collecting nectar from returning older foraging workers. As the 12-day-plus workers transport the nectar, they continue to process it and pass it to other bees to place in the comb for storage. At this stage, this nectar will still have too high a water content for storage, so workers will rapidly fan their wings to increase evaporation and reduce the moisture. Once the cells in the comb are full, and the nectar is less than around 20 per cent water (and can now be defined as honey), the workers will cover it with a sealing layer of wax

Honey stores being capped with wax

Stores of yellow and orange pollen

for long-term storage. As part of their food collection activities, the workers will also take pollen from forager bees and pack it into the cells of comb for long-term storage. During this process nectar is added, forming a protein-packed food source known as bee bread.

Guiding other bees home

An interesting further action that the workers are frequently seen carrying out is fanning their Nasonov's gland. This gland is at the end of the bees' abdomen and secretes a pheromone that attracts other bees to the colony/queen. Particularly when the hive is disturbed, some bees will be seen facing the hive with their rear legs extended fully such that their abdomen is raised, and the gland is exposed. Their wings are seen moving very vigorously, which wafts the pheromone away from the hive – thus letting other bees know where the hive entrance is and guiding them home.

Guarding the hive

The next role for a worker is protecting the hive from intruders. This isn't only from pests such as wasps, but includes mammals such as mice, bears and beekeepers! Workers assigned to this task also need to stop 'robber bees' from other hives

Guard bees protecting the entrance to a hive

entering and stealing stored honey/nectar from the hive. The guard bees are very sensitive to the differing scents from invaders and will release their own alarm pheromones to signal to other bees on guard duty to attack.

Bees will attempt to repeatedly sting unwanted visitors sufficiently to deter their intrusion, and quite likely kill them. Once the bee has stung a victim it will try to pull away, but the barbs of the sting cause it to detach from the bee along with the venom sac; the venom continues to be pumped into the victim until the sac is completely empty. Due to this damage, the bee will die, while the intruder has an extremely painful experience.

Foraging for food and other resources

This is the final role for most workers. They will collect pollen, nectar, water and/or propolis resin, preferably from close to the colony, but up to a 3-mile radius from the hive. Nectar is carried by the bee in its dedicated honey stomach, where it undergoes changes, prior to regurgitation and further processing in the hive by other bees to make honey (see p.140).

Water is collected only when needed to sustain the bees and for use in cooling the hive by depositing on the comb and being fanned to achieve evaporative cooling. Water is also carried by the foragers in the honey stomach.

Pollen is carried in the pollen basket or corbicula, which is an area of long hairs surrounding a portion of the rear legs that hold the compacted pollen. The components to create propolis are also carried in the corbicula, and consist of a mixture of plant-sap-like secretions, that are modified in the hive to form propolis. Typically dark brown in colour, propolis has antifungal and antibacterial properties, and is used to seal cracks and gaps in the hive, and to encase foreign objects that cannot be removed (see pp.164–5).

Worker bees' sensory capabilities

Foraging is only carried out by worker bees and provides a fascinating insight into their sensory capacities and communication behaviours. As you may have spotted in the anatomy diagram, and compared to us mammals, the honeybee has an unusual array of eyes.

It has a total of five eyes, two of which are compound eyes, and the three others are known as ocelli. The ocelli are located towards the top of the head and are simple, single lens eyes that are used as light sensors when flying to understand the location of the horizon and the position of the sun.

The compound eyes are much larger (even bigger on drones) and have multiple lenses that give a wide-angled view of the world around them. Their ability to detect polarised light helps with navigation relative to the sun's location, and high sensitivity to movement aids a rapid reaction to predators etc.

The forager is attracted to a flower based on its colour, shape, odour and proximity to the colony. Bees see colours differently to humans, and cannot see the red end of the spectrum, although they can see far into the blue end – notably, and unlike humans, they can see ultraviolet. It is interesting that many of the flowers that are most attractive to bees have a blue colour, such as lavender, borage, rosemary, lungwort and Viper's Bugloss.

Invisible to us humans, and to encourage pollination, some flowers have evolved target-like patterns visible in the UV spectrum for the bee to aim for as it tries to collect nectar and pollen. For example, the all-yellow (to us) dandelion flower has a pattern in the centre to indicate where to find nectar.

Communicating through dance

Once a foraging worker bee has detected a valuable source of food, it has an exceptionally clever way of letting other foragers know where it is – through the medium of dance! There are various dances that they may use, the most descriptive being the waggle dance, which lets other bees know the direction to the food source, and the distance to it (actually, the energy requirement to get there – which takes account of wind speed!). Bear in mind that this communication occurs in almost complete darkness, and on the vertical beeswax comb.

The forager bee's 'waggle dance'

The bee will move in almost a figure-of-eight, but with a vigorous shaking of its abdomen at the point the two circles intersect. The duration of this waggle indicates the distance to the food (1 second is approximately 1km). The angle of the waggling, relative to vertical, is the direction to the food source. This direction is based on the position of the sun in the sky, so if the waggle is 30° to the right of vertical, then 30° to the right of the sun's position is the direction they must travel to the food source.

Even if the dancing bee has been in the hive for some time after finding the food source, her dance will be modified to take account of the movement of the sun. The vigour of the waggle is also thought to be an indication of the quality of the food source. This strategy even works well on a cloudy day, as the workers' compound eyes can recognise the patterns of polarised light that pass through the clouds originating from the sun.

There are several other less complex dances that foragers are known to carry out, including the round dance, which relates to a food source that is closer to the colony. Other versions are thought to indicate warnings, or even preparations for swarming.

Worker bee population in the hive

The number of worker bees in the colony varies dramatically through the year, to align with the best summertime periods of foraging – thus ensuring there is enough food to survive through the winter and other times of low resources. A hived colony can store 2 to 3 times the quantity of honey it needs to survive the year. Beekeepers exploit this, but care is needed not to over-enthusiastically remove too much.

The lifespan of the worker bee depends on the time of year. In the busiest period of foraging in the summer, workers will work continuously up to their death after just 5 or 6 weeks. However, in the less active period over winter, when they are primarily striving to keep the queen warm to ensure her survival, they can live for 3 to 5 months.

The life of a drone

Drones form a relatively small population in a hive. Over winter there are none at all, but during the summer mating season there can be thousands in a colony. In a managed hive drones may form 5–10 per cent of the population.

Their creation starts with a queen laying an unfertilised egg in a purpose-built cell in the comb that is wider than the cell of a worker egg. It is thought that the extra width is detected by the queen, which results in the sperm she is storing not being used in fertilisation.

This drone egg will hatch after 3 days and is then fed royal jelly for the next 3 days. Thereafter its diet consists of bee bread (a stored pollen and nectar mix) until the cell is capped 7 days after hatching. The larva then pupates for 14 days, emerging from the cell on day 24 after the egg was laid.

It is notable that the drones have the longest period from egg to emergence of all the castes. The capped cells of drones can easily be identified as the cap is quite strongly domed – significantly larger than a capped worker cell.

The drones have only one job – and that is to mate with a virgin queen on her mating flight. They do not forage, clean the hive, build comb or feed the colony. They are initially reliant on being fed by the workers, but later just take the stored honey as required. However, this apparently charmed life has some drawbacks.

The process of mating is quite remarkable. Drones from many colonies, wild or from apiaries, will leave the hive and head for a drone congregation area where they will wait for an unmated queen to arrive, periodically returning to and from the hive to feed. These congregation areas can be anywhere between 1km and 5km from the colony and are situated in the open, away from trees and at least 5 metres above the ground (or up to 30 metres). The virgin queen will fly through the congregation area at height, thus ensuring only the stronger drones will be able to pass on their genetic material. The chasing drones are attracted through sight and smell of the queen's pheromones, and many will follow. If a drone catches up

Drone brood with a domed cap

Drone bee
(carrying a varroa mite)

with the queen, he will grip onto her and endeavour to mate. If successful, his endophallus is torn from his body and he falls to the ground, dying shortly after.

The queen will then continue flying in this area with the group of drones, repeating the mating process typically 15 to 20 times (although this may be over several mating flights). The queen will return to the hive after successfully mating, storing the collected sperm from the various drones for the rest of her life in her spermatheca, and never again repeating her mating flight.

Drones that have not mated also return to the hive at the end of the day, revisiting the drone congregation areas whenever possible the next day. If they do not die due to mating, or of old age, they will be forced from the hive at the end of the summer by the workers (to ensure they do not use valuable stores over winter) and will die of starvation. Their life expectancy is up to about 3 months.

One further complication with the production of drones is that some worker bees can be capable of laying eggs. As this laying worker has never mated, she will only create drones. It is thought this may be a continuous process, but it can be problematic for a beekeeper if eggs are seen in a queen-less hive and it is assumed she is still present. However, this can be identified by the beekeeper as the laying worker tends to lay several eggs in a cell, and they are sometimes on the sides of the cell as the worker has a relatively short abdomen.

Typical queen bee

The queen bee

The creation of a queen is an amazing process. Several days after a fertilised egg has hatched, worker bees decide if the larva is destined to become a queen. If so, the larva diet will continue to be royal jelly, and not change to include pollen and honey (as it does with workers and drones). The solely royal jelly diet activates the genetic switch that causes the larva's reproductive system to develop; it also modifies other physical attributes that make her differ from a worker. (This is an interesting area of biology known as epigenetics, where changes in gene function occur without changes in the DNA sequence.)

Once the workers have extended the feeding of royal jelly they will create and/or extend the uniquely shaped cell that the larva will grow and pupate in. The completed queen cell looks like a peanut shell attached vertically on the comb, although it is actually a long extension to a regular cell in the comb. The cell's relatively large volume allows the larva to be bathed in large quantities of royal jelly and grow very rapidly, such that she passes from larva to pupa to emerging bee in less time than a worker or a drone. (From egg to emergence is around 16 days, compared to 21 for a worker, and 24 for a drone.)

There are several drivers for the colony to decide to create a new queen (see overleaf), and it is potentially catastrophic for its survival if the process is not a success. Therefore, normally several, or many, queen cells are created by the workers – giving an increased chance of success. In this case, the first queen to emerge will pass through the hive killing any other developing queens in their closed cells by using her sting, after chewing through the cell wall, and/or battling to the death with any other emerged queens. The queen does not have a barbed stinger, so repeated use of it will not kill her – unlike the workers.

A few days after emerging, the virgin queen will go out on her mating flight, and will start laying eggs several days after returning to the hive.

There is approximately a seven-day window of opportunity for the workers to identify and react to a missing queen, as a larva on day 7 may have already commenced the transformation to a worker due its change away from a royal jelly diet. Even if this larva reverts back to being fed royal jelly, and grows as a queen, she may not be fully functional/fertile.

If a fertile queen cannot be created the colony is doomed. (Unless an observant beekeeper intervenes, see pp.112–5.)

Once egg laying has started, the queen provides a cohesion to the colony through her release of pheromones (often referred to as the queen substance), which permeate through the hive and are detected by all the bees. She endeavours to ensure the survival and productivity of the colony for several years and can be remarkably robust. Clearly, though, the colony needs a strategy to keep her alive to lay eggs, while being able to rapidly create a replacement if necessary.

When a queen is observed, she is often surrounded by a small group of worker bees. These are known as her retinue and they fulfil her needs by feeding her (with pre-digested food) and grooming. These interactions, and the continual replacement of the retinue members, distribute her pheromones efficiently around the hive.

Egg laying

The queen's normal routine is to progress around the hive looking for empty sections of comb to lay eggs in, lowering her abdomen to the base of the cell and depositing an egg. At her peak, when preparing the colony for its maximum worker population during the summer nectar flow, she will have her highest laying rate of 1,500 or more eggs per day.

Often, the queen will lay in a circular pattern on a hive frame, starting at the centre and spiralling outward towards the edges (pictured on pp.56–7). As this process may take several days to complete, it is common to see frames of eggs and brood containing circular patterns at all stages of development. This sequence is repeated and reinforced as the queen starts to lay eggs again in the newly emptied central cells where the new bees have emerged first.

It is important for a beekeeper to be able to recognise eggs, but this can be quite a test of your eyesight. (If you have reading glasses, it is best to put them on when

Capped 'swarm' queen cell at the base of a frame, with two partially built ones

you peer into a hive.) The eggs look like very small grains of rice, but they are only about 1.5mm long and up to 0.4mm in diameter. When they have just been laid they project vertically from the bottom of a cell, but recline to a horizontal over the following 2 days, prior to hatching into the initial larval stage.

Following the peak laying season the queen will slow down egg production, eventually coming to a stop for periods of the autumn and winter, as temperatures fall. The workers will form a cluster around the queen in these coldest months, ensuring she is kept warm enough to survive until the next season, when she will gradually increase laying as the temperatures warm.

Triggers for the creation of a new queen

There are a number of factors that can trigger workers to decide they need to start creating a new queen:

- **The queen dies** In this case the workers will notice a rapid drop in her pheromones in the colony. They will immediately start to make new queen cells. This is known as supersedure, and the queen cells created will typically be towards the centre of the comb, and not particularly numerous.

- **The queen's fertility is low** This may occur through old age, illness and/or depletion of her store of sperm. The workers will recognise she needs replacing, again through a fall in pheromone levels. They will start to make queen cells even if she is still alive. Again, supersedure queen cells will be built by the workers. A new queen will be created, and she may even tolerate the presence of the old queen.

- **Overcrowding** If the hive space is insufficient for the quantity of bees, brood and stores, the workers will decide to swarm. Again, queen cells are created; these are known as swarm cells and typically located towards the base of the comb – ten or more may be found.

  Around the time the newly created queen cells are capped, swarming appears to be imminent, and there will be a rapid departure of approximately half of the workers in the colony along with the old queen. They may come to rest temporarily (for hours to a couple of days) in a nearby hedge or tree in a large cluster, then later depart to a new target home. This may be anywhere they expect to be a safe place to start a colony again, such as a hole in a tree, a chimney stack or a bird box.

  Swarming is discussed in more detail on pp.106–10. Potentially it is an opportunity for a beekeeper to acquire free bees!

Keeping Honeybees

A beekeeper inspecting a busy hive

Many people are attracted to beekeeping with the idea of saving the bees and/or producing lots of free honey. It could be argued that neither of these aims is very realistic, but once you have started on the path of becoming a beekeeper, motivations continue to evolve and expand as the true nature of this incredibly addictive and fascinating hobby is recognised.

Looking after bees can provide a welcome distraction from other areas of our lives, and can be rewarding as we find ways to support these amazing creatures. As part of the learning process the new beekeeper develops an understanding of insect behaviour, pollination, biodiversity and seasonality.

For many, this leads to an expanding and enriching journey into the natural world – and a desire to protect it. This manifests itself in actions such as planting flowers that are good for all pollinators (bees, moths, butterflies etc.), avoiding the use of insecticides, caring and respecting nature's interactions, and encouraging biodiversity. More traditional neighbours may not initially approve of the path you have taken, but educating them about your new garden 'no mow' policy will become easier as they see the beauty of wildflowers and butterflies thriving – and receive a sweetener of an occasional jar of precious honey.

Despite occasionally being surrounded by frenzied bees, the focused, methodical and reflective demeanour required when handling bees creates (perhaps temporarily) an inner calm in many beekeepers. This can be further enhanced by the (fairly) common practice of talking to your bees – ideally when you are alone!

As our interest develops it is common to commence, or reinvigorate, other hobbies – many beekeepers enjoy learning new aspects of photography, carpentry, gardening and cookery. If you're up for it, there is even an option to pursue formal beekeeping qualifications through national organisations. In the UK, the British Beekeepers Association (BBKA) provides excellent support if you wish to follow this route. If you prefer a less formal approach, the local BBKA group will provide training, assistance and mentors to assist you on your journey. You will find that these colonies of beekeepers tend to be formed of friendly and, perhaps, somewhat eccentric individuals who love to share their knowledge and will welcome anyone – no matter their level of experience.

Of course, honey is a great product resulting from the bees' hard work (along with your interventions!). Arguably it is best eaten 'neat', but honey is also brilliant used as a flavoursome sweetener in a range of recipes (see pp.167–234), and to create drinks such as mead (see p.229).

Beeswax will also be available to you for potential use in numerous products such as candles, cosmetics, food wraps and polish etc. (see pp.236–243). You may find customers who are keen to buy these natural products, recognising that acquiring them from a local beekeeper offers a greater guarantee of quality and purity than highly processed and blended alternatives sold in supermarkets.

Bee hives

If you are going to keep bees you will, of course, need to invest in at least one hive and there are several options to consider, but first let's take a brief look at the history of bee hives.

Prior to the creation of artificial structures to house bees, humans collected honey by robbing from colonies in their natural locations such as voids in trees or rock faces. This opportunistic harvesting is depicted in some fantastic cave paintings at the Araña Cave near Valencia (see p.15).

The evolution of man-made hives

The first hive designs mimicked the natural void habitat of a colony of bees. This was achieved using clay pots or a woven structure (from grass, reeds etc.) known as a skep. In 2005, an archaeological dig in Israel found rows of 3,000-year-old clay pot hives, cleverly designed with a hole at one end for the bees to enter, and a lid at the other to allow the beekeeper to extract the comb.

Most pre-Victorian images of beekeeping depict the use of skeps. With these domed structures, management of the colonies was feasible. Honey and wax could be extracted by first driving the bees into another skep, or simply by removing small sections of comb. This comb could then be crushed to remove the honey. The colonies would tend to swarm naturally, allowing the beekeeper to collect and house a new colony – if they could safely catch the swarm!

There were some further evolutions to the skep design, aimed at avoiding the bees being killed and aiding the removal of honeycomb, such as the addition of wooden bars for the comb to be built from.

With a growing interest in natural beekeeping and less intensive honey production, there is increased enthusiasm for these historic techniques. Production of woven skeps is often demonstrated at farming and craft shows, and they can also be purchased from beekeeping equipment sites – although they are mostly used for collecting swarms prior to rehousing in a conventional hive.

Within the Victorian era there was a rapid evolution in hive design, with many patents filed. Notable highlights were the introduction in 1845 of a hive design by Jan Dzierzon, a Polish priest, with regular spaced top bars which is regarded as being the origin of many modern designs.

Within 10 years, the American Reverend L.L. Langstroth had created a similar hive that formalised the use of the bee space to separate hive components and, most notably, the frames. This important concept is regarded as a crucial component of hive designs. This is because if hive parts are spaced less than 7–10mm apart (roughly), the bees glue them together with propolis and wax; if spaced further apart, they will fill the gap with comb – causing similar problems and less tidy comb development on the frames. The bee space replicates the alignment of comb in wild colonies, allowing freedom of movement for the bees, while using the volume efficiently to maximise colony growth.

Although modified from the original design, the most commonly used hive variants across the world are general referred to as Langstroth (in particular in the USA). Most modern hives simply create the bee space between frames by having small, vertical sections built in that allow the frames to be self-spaced by pushing them together. These are still known as Hoffman frames after their creator, Julius Hoffman, introduced them in 1889.

In 1890 the WBC hive was created by William Broughton Carr. It became very popular due to its ease of manufacture. This attractive design is the one that most non-beekeepers picture as a typical hive, with its sloping roof and overlapping plank sides. It is effectively a small Langstroth-type hive that is housed within an external cover. This provides good insulation, which can be boosted further by filling with straw in the winter.

Jumping forward to the 1920s, the National hive was created as an inexpensive, movable hive and has become pretty much the UK standard hive. It is smaller than the traditional Langstroth, but beekeepers have found ways to increase the size if the bees need it.

A new colony creating comb inside a 'Freedom hive'

Hives at River Cottage

We have ten National hives at River Cottage (in two size variants) that are used in our teaching apiaries, and to provide honey for the kitchen and store.

We also have another four types of hive at River Cottage that are used to promote and demonstrate beekeeping with less focus on the production of honey, and with greater emphasis on providing a natural colony environment with less human intervention. These are the Freedom or Rocket hive, the Box Tree hive, the Top Bar hive and the Warré hive, and their designs form a continuum from solely a home for honeybees to thrive, to hives where a limited amount of honey can be collected. One thing they all have in common is that the bees build the comb naturally, without the use of foundation placed in frames. Consequently, any honey collected is naturally encased in wax entirely made by the colony. The honey can be used in the form of cut comb or pressed from the wax. It cannot be removed from the comb in conventional mechanical rotary extractors.

In order of potential honey productivity (low to high), our hives are as follows:

The Freedom (or Rocket) hive

Designed by Matt Somerville of Bee Kind Hives, this consists of a vertically orientated hollowed-out log with a thatched roof standing on three legs to raise it substantially above the ground. This closely reflects the natural habitat of honeybees in a tree hollow. It does not provide a means for a beekeeper to collect honey. Effectively, it can be seen as aiding the rewilding of honeybees.

There is a small access hatch at the base of the hive, which allows observation of the progress of the colony. Bees enter and exit through several holes drilled through the log. Once bees are populating this hive, it really doesn't need any intervention by the bee observer/beekeeper. (It should be noted that in some countries there are regional regulations that stipulate the use of hives with moveable frames only, which could preclude the use of this hive, or those of a similar design.)

The Box Tree hive

Designed by Matt Sommerville and Hugh Fearnley-Whittingstall, this can be strapped to a tree trunk and emulates the habitat created by the Freedom hive. It has a removable roof to allow later expansion by adding further capacity on top – potentially for the collection of comb honey. Externally, it looks like a tall rectangular box, but there is a vertical cylindrical internal hive body of similar dimensions to the Freedom hive (constructed from strips of timber in a similar fashion to a barrel), encased within the box. There is sawdust insulation between the two layers. Again, this is very similar to a 'wild' natural environment for the colony and needs minimal beekeeper intervention, but with the option for expansion.

Freedom (or Rocket) hive

Box Tree hive

Top Bar hive

Warré hive

Comb produced in a Top Bar hive

The Top Bar hive

This is an old design that can easily be made from scrap wood (designs are readily available from the internet), or bought from several beekeeping equipment suppliers. To the bees, it mimics a horizontal log, but it is almost triangular in cross-section.

The bees will create comb that hangs from wooden strips, called top bars, that we add to form the top of the living chamber. (We dribble some beeswax into a groove roughly cut into the bars to encourage a neatish starting point for the bees to build from.) These hives can be customised/upgraded by adding viewing windows and a hinged roof.

The bees enter the hive through holes drilled through the side wall. To aid the colony, the internal chamber size can be extended (or shrunk) by moving dividers placed between the top bars. If needed, it is possible to feed the bees by supplying sugar syrup using perforated jars placed within the hive.

The bees will tend to place stores of honey at one, or both, ends of the hive. This provides the opportunity for the beekeeper to remove some of these strips of comb to extract the honey.

Bars from a Warré hive with freshly drawn comb

The Warré hive

This was designed by Émile Warré in the first half of the twentieth century. It has visual similarities to the conventional National hive, in that it is constructed as a tower using separate boxes, but it uses wooden strips for the bees to build comb from (rather than frames) and has a smaller cross-sectional area.

As the colony expands, additional boxes can be added at the bottom, allowing the bees to build new areas of comb to expand the colony. Honey can be collected from this hive late in the season, by removing and processing the comb in the upper box of the hive – where the bees naturally tend to place food stores.

Adding a queen excluder to a National hive brood box

Adding a super with frames

Adding a crown board

Complete hive, with roof in place

The National hive

This is the most commonly used hive in the UK and the one you are most likely to encounter as a new beekeeper. A single hive is a great start for a new beekeeper, but if you become hooked on the hobby you will quickly wish to expand. If possible, it's beneficial to start with two hives and colonies.

Although a simple design, the National hive is fairly impractical to make yourself. However, a complete hive kit can be bought relatively cheaply from several suppliers (see Directory, p.250), particularly if purchased as a flat-pack starter kit that just needs nailing together. Most external components are made from cedar, so they last outdoors for many years; cheaper, less robust pine variants are available. The components of the hive are not locked together in any way, they just sit on top of each other, and can easily be separated and reconfigured when needed.

Starting at the bottom of a typical hive, the components are as follows:

Hive floor

This is a square frame with the hive entrance on one of its four sides. The entrance can be restricted in size by adding an entrance block. A smaller entrance is useful when the hive is weaker over winter, and allows easier defence of the colony at times when predators, such as wasps, are abundant at the end of summer. The bees have uninterrupted access to the hive entrance/exit from their nest in the section above.

The base of the floor was originally designed as a solid board, but now (in the varroa mite era) it is most commonly an open-mesh floor. This allows any varroa mites that fall off the bees to pass through the floor and not clamber back up to harass the bees. The floor also has a convenient slot below the mesh to allow a lightweight board to be slid in, which can help the beekeeper get an understanding of varroa levels. This can also give a good idea of where most of the bees are in the colony, due to collecting bits of wax and other debris fallen from their manoeuvres in the body of the hive. A convenient additional feature of the mesh is that it provides good ventilation to the hive, reducing the risk of condensation.

Brood box

This is where most of the interesting bee action occurs! The brood box sits on top of the floor, and is the home of the queen and all her eggs, larvae and sealed brood. The box usually contains 11 self-spacing (Hoffman) frames.

The queen will remain in the brood box throughout her life (after mating). She lays eggs in the cells in the comb in the frames, which will later emerge as fully formed bees. Thereafter she will use these cells to lay further eggs. The worker bees that form the majority of the colony will be bringing nectar and pollen into the brood box for storage or use. Often it's possible to see stores of pollen and honey with brood on the same frame.

Frames

A frame is a timber rectangle that is used to house the beeswax comb created by the bees. Lugs on the top bar of the frame are used to hang the frames within the brood box, and are the handholds for the beekeeper to use when examining the frames. Most beekeepers place a purpose-made sheet of beeswax within the frame that is known as the foundation. This has a shallow moulded comb pattern in it that the bees will draw out to form a very neat area of comb. The foundation is usually strengthened with very thin wire.

The frames can be bought pre-made, but most frugal beekeepers buy them in packs of parts and nail them together. There are two potential alignments for the set of frames – known as the 'warm way', and the 'cold way'. If frames are positioned parallel with the entrance in the hive floor, the alignment is warm, as any draughts through the entrance may be blocked by the frames. If aligned perpendicular to this (as pictured on p.84), the alignment is cold, as draughts could pass to the back of the hive.

In practice, beekeepers have their own preferences, and we don't think it makes a lot of difference to the colony. At River Cottage, we use the warm alignment, which positions you to stand directly behind the hive when inspecting the colony (see pp.99–103) and the furthest distance from the bees coming and going through the hive entrance.

Dummy board

This is an extra piece of equipment that we use in the brood box. It is a similar size to a frame but is solid and has no foundation. It is positioned as a 12th frame, at the end of the row of frames, and is used to shrink any remaining gap at the end of the brood box once all the frames are squeezed together, ensuring the bee space is neatly maintained throughout by the Hoffman spacers on the frames. It has an additional benefit of immediately providing plenty of space to manipulate frames once it is removed. Many beekeepers don't bother with this, and just leave an equidistant gap at each end of the group of frames. This can lead to the bees filling these areas with weird shaped areas of wax, which can be a bit of a nuisance, but it does allow the bees more freedom to raise brood at both ends of the hive.

Queen excluder

There's a clue in the name. This keeps the queen in the brood box, stopping her passing to the area above. Most beekeepers use the version known as a 'wire queen excluder', which is constructed from carefully spaced metal rods in a wooden frame. The rods are spaced wide enough to allow worker bees to pass through freely, but close enough to stop the larger queen (and drones) passing. It is best to use a queen excluder that has a wooden frame, which creates a bee space between the top of the frames in the brood box and the metal rods – such that all bees can freely walk around below it. (Best to avoid using the cheaper frameless galvanised or plastic versions, to reduce the risk of squishing bees.)

The aim is to ensure that eggs are not laid by the queen in the super box above, which is designed to be the honey store that the beekeeper steals from. It avoids contamination of this honey with faecal matter from the larvae, or even some of the eggs/larvae being incorporated into it.

Typically, the queen excluder is only used when there is an expectation of lots of nectar entering the hive, so it will be removed over winter.

Supers

These are the boxes full of frames that we hope the bees will fill with honey. Typically, they have ten frames in them. This compares to 11 frames in the brood box, despite being a similar area. Running with one less frame encourages the bees to make the wax cells a little deeper, giving an overall higher volume of honey.

Supers (and frames) are not as tall as a brood box – they are about two-thirds of the height. This makes them far easier to remove and carry when full, when they will weigh more than 10kg. Unless collecting the honey to sell as cut comb, the foundation used in the frames is wired in a similar way to brood frames. This strengthening ensures the frames can be spun in a mechanical rotary honey extractor without falling apart.

Once the nectar flow is underway the frames in the super will fill very quickly (we hope!), so several supers may be put in place. Usually two will be in use, but this can be increased to four or more, as long as they are filling with nectar and it is physically possible for the beekeeper to access and move them.

Generally, supers are not left on the hive through winter, and will be stored in the beekeeper's shed after the honey has been removed. The drawn comb on the frames is a valuable asset and will be used again in future years if it is in good, clean condition.

Crown board

This is a thin board that sits on top of the supers (or the brood box over winter). It forms an internal roof to the colony. If needed, it can also be used to dispense food supplies to the bees by adding syrup feeders, or chunks of sugar fondant above slots that are pre-cut into it.

An additional use is in removing the bees from the supers when it's time to take away the supers for honey extraction. In this case, the slots have a spring-loaded one-way valve added, called a Porter bee escape. The crown board is then placed below the supers that are to be taken away, and over 24 hours or so the bees will pass through the bee escapes into the brood box, and none will re-enter.

Roof

This tops off the hive. It is fitted with a thin galvanised steel sheet on top to protect the timber hive from the elements. Most beekeepers use flat roofs, as they are more practical to use (and cheaper), though more attractive gabled roofs can be bought. When taking a hive apart to check on the bees, the roof is usually inverted and placed on the ground, and the hive parts piled on top of it – this is less easy to do if the roof is gabled.

The roof has some small vents added that allow air to circulate vertically through the hive, to lessen the chance of condensation forming.

Hive stand

You don't want to place the hive directly on the ground, so the complete unit is placed on a hive stand of some sort. It doesn't need to be complex, just strong enough to support the considerable weight of a full hive in the summer, and tall enough to reduce the amount of bending needed when the beekeeper checks and manipulates the hive (We like to raise the hive floor about 50cm above the ground). Although not a necessity, it's nice to add a sloping landing board to the stand, so that the bees stop on this prior to walking in through the hive entrance. It gives the beekeeper a good opportunity to watch and contemplate their activity.

Equipment for National-type hives

This section covers the minimum needs for a single hive and follows with suggestions for additional items that make beekeeping easier. Several of the beekeeping supply companies will provide good-value beginners kits, including many of the following items, at a discount.

Colony of bees When you acquire some bees, they will likely be provided as a nucleus. This is a smallish colony on five or six frames that will have a fertile and laying overwintered queen, and frames that contain sealed and unsealed brood, pollen and stores of nectar/honey. Most new beekeepers will acquire these via a local beekeeping group for a reasonable fee, but colonies can be bought from commercial beekeepers (usually at a higher cost). Acquiring them locally has the benefit of knowing the bees are already used to the local flora, and you may well gain the assistance of a local mentor. Another option is to collect, or be given, a swarm. This has the benefit of potentially being free, but care is needed to check that the bees are not carrying diseases and that there is a fertile queen present.

Hive The more conventional National-type hive will consist of a floor, brood box, queen excluder, two supers, crown board, roof, 11 frames for the brood box and 2 x 10 frames for the supers.

Hive stand This should include straps to hold down the hive in windy conditions. Unless the hive will be located in a very sheltered area, it is also advisable to fix the stand to a partially buried post or two to prevent it toppling in strong winds.

Hive tool There are several shape variants available, and it doesn't matter which you choose provided the tool has one end capable of scraping wax/propolis and prising apart hive parts, and the other end can be used to gently lever apart frames within the hive. It is useful to own several hive tools, as it is easy to lose them!

Bee brush Used to gently brush bees from a frame when required. The bristles traditionally consist of fairly stiff horsehair, but synthetic alternatives exist.

Watertight container To fill with a solution of water and washing soda (sodium carbonate) to clean your hive tools after use.

Protective clothing To prevent bee stings, effective protection is essential. A full veiled bee suit is ideal, but a half-suit with a veil is sufficient if you wear thick, loose light-coloured trousers. We've had uncomfortable experiences with half suits married with dark, skinny jeans, as stings can pass through these.

Gloves are, of course, essential. Opt for light-coloured rather than dark gloves – these can look like a bear paw to a bee! Beekeeping leather gloves are very tough but are rather cumbersome and difficult to wash. Thicker rubber dishwashing gloves work well too. We use thin latex or nylon surgical-type gloves to improve dexterity – but it is very easy to be stung through them.

Wellington boots are the ideal footwear, with trousers tucked inside them – this avoids some nastier stings, as bees like to climb upwards into dark areas!

Smoker and fuel Lots of types and sizes of smoker are available. The simple designs with bellows work very well, once you get the hang of lighting them. Organic fuels such as cardboard, wood chips, hessian and pine cones are the best choices, to minimise the risk of harming the bees or yourself. Larger smokers are best if you have a few hives to look at, or a distance to walk between hives. A gas lighter and a back-up of matches are needed to light the fuel.

Feeder There are occasions in the year when you may need to feed the bees to ensure they have enough sustenance to survive. This may be over winter, during prolonged periods of bad weather, or even during the summer when there is a gap in forage availability. A circular rapid feeder works well and is very cheap. It is placed on the crown board, over one of the access holes, and filled with a solution of sugar syrup.

Eke This is a simple square frame that is the same dimensions as the hive and is used to house the feeder.

Hive logbook Or hive record sheets to record the results of hive inspections.

Grot box A container to deposit any debris scraped from the hive during an inspection, such as propolis, dead bees, wax scrapings etc. This can then be taken away and disposed of separately (best not to drop on the floor by the hive as it may encourage pests). An empty sealable plastic ice cream tub is ideal.

Mobile phone If working alone, it's useful to carry one, in case of problems.

Antihistamine tablets For some people, these may be worthwhile to reduce the effects of a sting, or help with hay fever/allergies. The aesthetics of sneezing when wearing a beekeeper's veil are not great.

Honey extractor These can be expensive, but it is often possible to hire one from a local beekeeping group (book early, as they are in high demand at the end of summer). If you decide to buy one, the capacity and type (automatic or manual), will be determined by the number of productive hives you have, and your budget.

Food-safe buckets with lids To store honey in. You need at least one bucket with a tap so that the settled and filtered honey can easily be decanted into jars.

Stainless steel double strainers To clear pieces of wax (and bits of bee) from your extracted honey prior to storage.

Jars Clean and sterilised jars and lids for storing/gifting/selling your honey.

Potential additional equipment

Additional hives Useful as spare equipment and to help with managing the hives.

Nucleus box(es) With frames and foundation, these are extremely useful for many purposes, including swarm control and collection, breeding new colonies, queen rearing etc. They can also function as a good standby small hive that a colony can live in for long periods.

Brood boxes and supers Extras, with frames and foundation, are needed in case any components of the hive are damaged, for swarm control and if your hive needs to be expanded. They are most useful if they already have drawn comb in them.

Clearer boards To persuade bees to leave the supers prior to extracting the honey. You can use the crown board fitted with Porter bee escapes (often supplied with a new hive), but it is faster if you use boards fitted with a rhombus bee escape.

Queen marking kit This includes a cage bottle (to catch and hold the queen while you mark her) and different coloured marker pens (see p.97). As time goes by, you will want to mark new queens to aid identification and their year of birth.

Varroa treatments There are several chemical treatments and natural remedies available, as well as items of hive equipment (see pp.119–20). Most beekeepers will treat the colony in autumn or winter.

Fondant An additional food source used in winter that is left on the hives for the bees to use if necessary. Simple bakers' fondant (only sugar and glycerine) works well, but tailored varieties are available that contain protein and other nutrients.

Refractometer This tool rapidly confirms the moisture level of honey and nectar.

Uncapping tool There are various designs you can experiment with to aid the removal of the capping wax seal in order to access the honey prior to extraction, but a breadknife also works well.

Uncapping box Although a stainless-steel oven tray works well, you may wish to upgrade to purpose-designed trays/boxes when uncapping the frames and filtering the honey sliced off with it.

Travel screen This can be used as a lightweight and ventilated mesh cover, replacing the crown board and roof, to ensure the colony does not overheat (or escape) when being transported between sites.

Pollen colour chart A paper chart or phone app that helps you identify the pollen stored in your hive by colour and season – based on typical flowering plants.

Equipment for other hives

If you decide to go for the less productive, but lower intervention hives such as the Freedom, Warré or Top Bar, you'll require less equipment. You don't really need any of the items described above with the Freedom hive, as you will only be observing it from a distance (and not collecting honey). For the others you'll want to check on the hives, so protective clothing and smoker are needed, and if small amounts of honey are taken, you will likely need a honey press for extraction and buckets etc.

A typical apiary

Choosing an apiary site

Once you have decided to keep bees, you need to figure out where to house them. There are a few crucial considerations when selecting a site, and several others that are not mission critical, but will improve your chance of success.

Firstly, you must choose a site where there is sufficient forage/resources for the bees: essentially nectar, pollen and water. Nearby hedgerows, woodland and gardens can be beneficial due to the diverse range of flowering plants that grow through the year. Towns and cities can also be suitable, due to abundant flowers in gardens and parks, but you will only be able to place a few hives in an average suburban garden.

A level site is important, such that the hives are not sloping. Terracing the ground to make the hives level is fine but can be a lot of work. Avoid damp, low-lying areas that may also become frost pockets, or even flood. You will also need to keep the hives protected from strong winds. If the best site available is exposed, a wind barrier near the hive may be beneficial.

You'll need sufficient space to work behind the hives, and to allow for expansion. Not only will your enthusiasm likely result in more hives, but you may also need space to house swarms you collect. Small gardens can be problematic, as a hive can be very active on a warm day, causing family and neighbours to evacuate the area!

Avoid a location where the bees' flight path (the beeline) will be close to people and/or livestock. Next to a public footpath/bridleway, or pointing the bees at your neighbours, won't be welcomed! If you put thick fencing or hedging that is 2 metres or more tall, and a couple of metres away from the hive entrance, it will direct most bees above head height, but you can't rely on this next to a busy pathway. Fencing the area off is important to discourage people and livestock getting too close.

Easy access is needed to allow the positioning of the hives and the addition or removal of equipment. It's great if you can take your car close to the apiary. Remember, a full super will weigh more than 10kg. Carrying one some distance will soon cause an aching back, and increase the risk of a fall. Similarly, a hive stand of a suitable height for the beekeeper improves the ergonomics significantly. You also don't want to be climbing over fences, or tripping over obstacles – particularly if you need to make a rapid exit.

Other factors to consider

If possible, it's best to avoid being visible to the public. Interested passers-by may be tempted to come over and look too closely, which is frustrating when the hive is open during an inspection. And, unfortunately, theft and vandalism are not unknown – try to be fairly secretive with your location!

It's also preferable to avoid placing hives beneath trees as this can discourage the bees from flying if sunshine is restricted, and they may be irritated when rainwater is dripping from the trees. It's great if the hive gets sun on the front early in the day as it encourages the bees to start their day's foraging, but it isn't crucial.

You really do not want to be regarded as a nuisance at your chosen location. Taking care to minimise swarming is wise and the occasional donation of a jar of honey to neighbours is usually a welcome strategic move to maintain good relations.

Setting up a hive

It can seem quite daunting setting up your first hive, but fortunately it's a fairly simple process and there is not a lot can go wrong. You just need to have a plan you can be confident in, and perhaps even rehearse the actions.

It's important to first reflect on a few aspects. The best time of year to install a colony is from mid-spring to early summer when ambient temperatures should be increasing, the queen will have commenced laying and the colony will be growing. With this early-season start there is also a good chance that the colony will expand rapidly into the nectar-flow period, creating enough honey stores for you to take some – quite a thrill in the first year.

That said, it is entirely feasible to be successful for most of the period from April to August – as long as you acquire a strong and fertile colony and the weather is kind. A late August start may be borderline as colonies have started to shrink, forage is declining and pests such as wasps will be predating on weak colonies. It's best to avoid autumn and winter.

Getting some beekeeping training over winter, and acquiring bee-handling skills in early spring is strongly recommended – this can often be achieved through your local BBKA group. Furthermore, you will have had plenty of time to obtain the necessary equipment (we recommend having an extensive Christmas wish list) and early season supplier sales are often held in January.

Once you have this equipment, a source of bees and an apiary site, it's time for the really exciting part: setting up the hive.

The ideal scenario is if you have a mentor (maybe from the local BBKA group, or even the person providing the bees). No need to worry if this isn't feasible, you just need to be more systematic and meticulous with the procedure.

You will most likely receive your first bees in the form of a nucleus, which is a small transportable hive that holds about six frames of bees, so we'll focus on this. The colony should be growing and include frames with eggs, larvae, sealed brood and food stores. It should be disease-free, and composed of mostly worker bees, a few drones and a single laying queen. The frames in this colony are going to be transferred directly to your hive, so they need to be the appropriate dimensions for the hive you have acquired.

Step-by-step guide to setting up a hive

- Choose a suitable day of weather for this activity: the ideal scenario is sunshine, warm temperatures (higher than 16°C), low winds and no rain. This avoids the bees being shocked by rapid environmental changes, and makes the whole experience more pleasant for all. (You can carry the work out if the weather is worse if you are constrained to do so, but avoid it if you can.)
- Be ready for beekeeping activities, wearing your zipped-up suit, with hive tool ready and your smoker running.
- You need to have the new hive stand in position, ensuring it is stable with sufficient space behind for you to stand and inspect the hive. You need to align the hive entrance in the direction you want the bees to emerge on their flights. Often it is recommended that the new colony is placed in position on your hive stand in the original nucleus box to acclimatise to the new environment for a day or two. This may be regarded as a gentler approach, but we find this is not really necessary and feel it is best to complete the move into the new hive in one event.
- Place your new hive on the hive stand, with the roof and crown board removed, and the brood box empty of frames. The frames should be placed close by. For a National hive you will likely only need five frames, as there should be six from the nucleus, giving you eleven in total in the brood box. There is no need to use any supers or a queen excluder at this stage.
- Gently place the nucleus of bees (entrance closed) next to your new hive. Leave it alone for a few minutes while the colony calms after being moved about.
- Now it's time to start interacting with the bees! Lift the roof from the nucleus box and calm the bees with a gentle puff from your smoker.
- Next, lift out the frames gently, but fairly swiftly, and place centrally within the new brood box in the same sequence as in the original nucleus. It is beneficial to lift the frames in pairs, as this creates less disturbance to the colony. Don't linger and inspect each frame – but keep a look out in passing for the queen (hopefully she's marked) as it can offer assurance that all is well.
- Once all are in position, place your own extra frames either side of the installed ones, and push them tightly together so they are neatly self-spaced due to the protrusions on the frame. You may now find many bees are left in the nucleus box or returning to it (despite it being empty of frames). If so, invert the box over the new hive and then gently brush the bees out. If necessary, a short tap on the nucleus will make any remaining bees fall out.
- Replace the lid on the nucleus box, so that flying bees don't return to it, and place to one side. Finally, replace the crown board and roof back on the new hive, taking care not to squish any bees.

- You can now have a rest and watch as the bees recognise the queen and colony are now in a new home. You will see several, or many, worker bees outside the hive aligning themselves, radiating from the entrance of the hive and fanning their Nasonov's gland to indicate to others that this is where the colony is housed. Flying bees will gradually enter the hive once they recognise the colony is safe to enter.
- If you found that the nucleus frames had little weight of nectar and honey stores, and particularly if undrawn comb was added, it is worth feeding the bees via a rapid feeder placed over the crown board and filled with a solution of 2:1 sugar:water. This will provide a very useful and accessible source of food for the colony, and is likely to encourage them to rapidly adapt to their new environment and draw new comb. You will need to add an eke, or an empty super, to surround the feeder, prior to adding the roof.

Although virtually impossible to resist, it's best not to disturb the bees for a few days – so try to avoid opening the hive and inspecting the frames. In the meantime, it's acceptable to quietly visit and watch the hive and bees. If you see pollen arriving at the entrance, it's a positive sign that the bees have started foraging and are adapting to their new environment. If you can resist leaving a full inspection for 4 days, any eggs seen will have been laid after adding the bees, indicating the queen has been in residence since the colony was installed, and has adapted well to her new home.

One final consideration: if you acquired the colony of bees from another beekeeper, conventionally you would give them 6 frames and foundation to replace the ones they supplied in the nucleus.

You should now initiate your sequence of weekly hive inspections – this is a great opportunity to identify, note and track the development of your new colony.

Inspecting hives

This is the best part of beekeeping. It's an opportunity to interact closely with the bees, and see in detail how the colony operates. When the bees are calm, it is a rewarding learning experience for the beekeeper. You can observe behaviours we read about – such as bee dances, fanning, pollen delivery to cells, the queen laying eggs. You can gain a snapshot understanding of the nature of an individual colony in terms of behaviour, health, progress and sustainability.

It's important to be relaxed and methodical, and to tune in your observational skills when you inspect the hive. If all goes well, you'll hopefully develop an inner calm during the process.

Unfortunately, a hive inspection is an immense disruption to the colony, and it is clear the bees are averse to it. Opening the hive conflicts with the idea of low intervention beekeeping, but, if you want to ensure the hive will be productive, it can be regarded as a necessity.

When looking in a hive it can be very difficult to find the queen due to the quantity of workers wandering across the frames, and the chunky drones seeming to imitate a queen – it's easy to hide within a family of 50,000. Nevertheless, it's very useful for the beekeeper to be able to find the queen when inspecting the hive, as it allows the opportunity to do some advanced beekeeping trickery, such as creating additional colonies or trying to avoid swarming, or simply offering some reassurance that all is well.

To help with the search it is a common practice for the beekeeper to put a small blob of harmless coloured paint on the thorax of the queen. Even with this paintwork, there is no guarantee she will be seen, but it certainly increases the likelihood. The five colours used relate to the last number in the year of creation of the queen within a 5-year cycle (as a queen will rarely survive for more than 5 years). The mnemonic 'Will You Raise Good Bees' is a convenient way to remember the cycle, as shown below.

The conventional colour coding system for marking queens

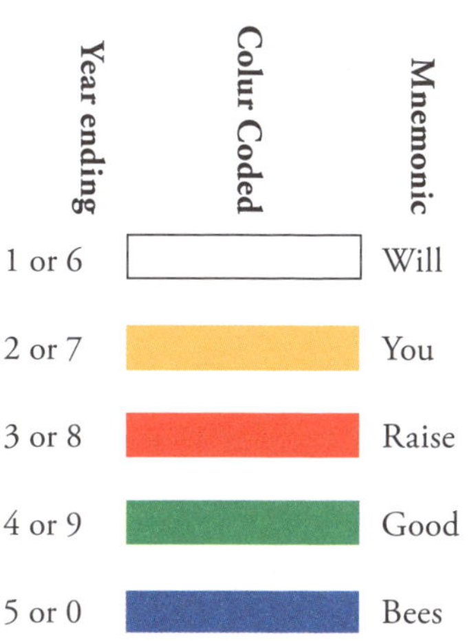

Year ending	Colur Coded	Mnemonic
1 or 6	(white)	Will
2 or 7	(yellow)	You
3 or 8	(red)	Raise
4 or 9	(green)	Good
5 or 0	(blue)	Bees

For example, a queen that emerged in 2025 will be painted blue.

A queen marked green, 'created' in 2024

A beekeeper carefully inspecting a brood frame

Inspection procedure

It is important for the beekeeper to be respectful to the bees, i.e. be gentle, and responsive to their temperament. A gentle puff of smoke into a hive is very effective at reducing the likelihood of receiving stings from a colony. The smoke is used to suppress the bees' sensitivity and response to their alarm pheromones – secreted by guard bees that are trying to galvanise a wider defensive response. When opening a hive, we always have a smoker lit and at hand, but try to minimise smoker use and focus on responding only to observed defensive/aggressive behaviour.

Lighting the smoker is simple, and just requires dry organic materials and an ignition source. A reliable method is to light a small amount of shredded cardboard, place it at the bottom of the smoker, then add sawdust until the smoker chamber is full, while simultaneously gently squeezing the bellows to heat the sawdust. The lid can then be hinged shut, and the bellows compressed vigorously until cool grey smoke is streaming out. There should be no visible flames and little heat. The smoker will continue to smoulder for an hour or more if puffed intermittently.

It is best to complete the inspection as quickly as possible, but keep at a speed that allows you to be thorough, and avoid knocking hive parts together. You will quickly learn that sudden movements cause bees to rise to check what is happening.

When putting a hive back together, lower each section slowly to minimise vibration, and always initially place the piece down such that it is rotated horizontally about 30° relative to the part below. There will now be a lot less contact area between the two sections (only eight small contact points), minimising the area to squish bees! The added part can then be rotated slowly into position, gently pushing bees out of the way (maybe aided with a puff of smoke).

The inspection procedure differs slightly according to the number of supers and brood boxes in place on the hive. The simplest situation is early (or late) in the season when there may be no supers on the hive. In this case, on the hive stand, you will simply have a floor, with a brood box above, then a crown board, then a roof.

Step-by-step guide to hive inspection

Prior to inspection you will need to have your smoker running, a hive tool at the ready and a means to record hive notes.

- Without opening the hive, view the bees coming and going from the entrance. It's reassuring if you see a continuous stream of bees leaving, and others returning with and without pollen. Watch out for undesirables, such as wasps, entering the hive. Look for any apparent fighting between bees at the entrance, which could imply the hive is being robbed of stores by another colony. Be careful not to stand in front of the hive entrance, as this will cause a traffic jam of bees waiting to enter.

Timing of key stages of bee development

Day	Worker	Drone	Queen	Event
1				Freshly laid eggs, close to vertical at bottom of cell
3				Egg lying flat at base of cell and about to hatch
5		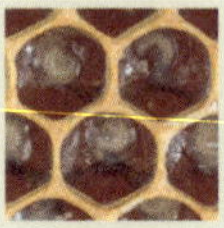		Young hatched larvae Queen cell being drawn out vertically
7	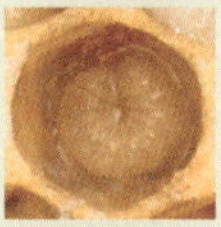	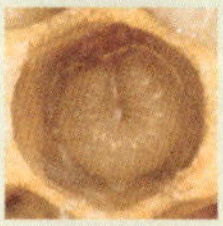		Larvae growing Queen cell being extended before capping
8				Grown larvae prior to capping Queen cell capped
9		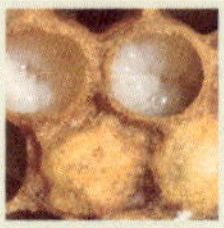		Worker cells capped Drone cells being capped
10				Drone cells capped
16				Queen emerged
21				Worker emerges
24				Drone emerges

- If you do not know the colony temperament well, then puff some smoke into the hive entrance, and perhaps under the open mesh floor of the hive. Observe for a couple of minutes, while the smoke takes effect. Use this time to look through previous hive notes, making a mental note of any specific observations that need further consideration during this inspection.
- Lift the roof from the hive, invert and place it on the ground, or on the hive stand if there is space. (It is important to invert the roof, so that the internal parts of the hive are not in contact with the ground, avoiding any creatures taking the opportunity to contaminate the hive.)
- With the hive entrance assigned as the front, you should now stand behind the hive if frames are orientated 'the warm way' or to the side of the hive if aligned 'the cold way' (see p.86).
- Remove the crown board. You will need to break any propolis seal that the bees have created between the crown board and the brood box. This is best achieved by placing the hive tool at a corner between the brood box and crown board and levering the board upwards. Then slide the hive tool around the edges to release it. Place the crown board on the upturned hive roof.
- You will now be looking at the tops of the frames of bees. Observe, and make note of how many seams of bees are visible, i.e. how many frames are occupied with bees. Early in the season there may only be two or three, later on there may be ten. Recording this in your hive notes gives you useful data to identify if the colony size is expanding or contracting over time. You may wish to puff some smoke across the frames if the bees are being aggressive and/or many are flying out from the top of the hive.
- If using a dummy board, place your hive tool vertically in the space between the first frame and dummy board and twist it to prise them apart. Place the dummy board on the hive stand, or on the roof to the side.
- You can now start an inspection of the frames. Use your hive tool to lever the first frame away from the second frame. If it is stuck in place by wax and propolis you may need to do this on both sides of the frame. Always lever at the end of the frame (at the lugs) to give maximum leverage, and avoid damage to the comb. This frame should now be completely detached from the next one.
- Gently lift this frame out vertically, ensuring that you do not rub the frame against the adjacent one and squish or roll any bees. Inspect this frame on both sides, noting the following:
    - Is the foundation drawn out into cells?
    - What do the cells contain – nothing, pollen, nectar, sealed stores, eggs, larvae, sealed brood?
    - Is there any sealed drone brood or worker brood?

- Are there queen cells?
- Any sign of disease?
- Are there drones and/or workers present?
- Is the queen on the frame?

- Once this first frame inspection is complete, the frame can be gently placed back in the hive, but flush against the side/end wall of the hive, such that there is a sufficient gap for the next frame in line to be removed. If a dummy board was not used in the hive, there may not be quite enough space to remove the next frame easily. In this case it is OK to place the first frame leaning vertically onto the side of the hive, or across the depth of the hive stand. It is best to be confident the queen is not on this frame, as we don't want her to wander off – if you find her, let her walk from the frame back into the hive.
- Next you will loosen, remove and inspect the next frame. Once complete, place back in the hive, but pushed tightly against the first frame that has already been inspected. This leaves a suitable gap to remove the third frame, and also ensures the already inspected frames are placed in a neat block without gaps that could lead to squashed bees when the frames are pushed together at the end of the inspection.
- Continue inspecting until every frame has been examined, and returned to the hive. Take special care if the queen is found on a frame (you won't always spot her). This frame should be particularly gently replaced, keeping an eye on her location.
- Once all frames are back in the hive, they will be in a single block that is about 2cm away from their original location. If a dummy board is in use, the whole block of frames can be levered tightly back to the original location at the opposite wall of the brood box, using the hive tool. The dummy board can then be returned and pushed tightly against frame 1.
- If a dummy board is not in use, the block of frames is pushed such that it is centralised in the brood box, with an equal gap at either end – remembering, of course, to return the first frame to the hive if previously removed.
- Place the crown board back onto the brood box.
- Invert and place the roof back on the hive. Quickly inspect the outside of the hive to ensure the boxes are neatly aligned, and bees are freely passing through the hive entrance.
- It is important now to complete the hive notes, to make sure you don't forget any details. Key information relates to the presence of a laying queen, healthy brood and the presence of food stores. Remember, if you don't notice a queen, but see eggs, then it is very likely she was there. In any case, she was certainly there within the last 3 days and you didn't lose her at your last

inspection a week or so before! Make sure you highlight anything in your notes that you may need to check on your next inspection, and/or any actions that need to be completed.

Inspection adjustments if supers are in place

If the hive has one or more supers in place, the sequence is modified a little. Let's assume you have two supers in place, and a queen excluder (to stop the queen laying eggs in this section that you wish to recover honey from). So, starting at the bottom of the hive, you will have a sequence of a floor, brood box, queen excluder, two supers, crown board, roof.

Proceed with your inspection but leave the crown board in place. Try to squeeze the hive tool between the queen excluder and the lower super at a corner. While applying some upward pressure on the two supers with one hand, try to crack the joint between the queen excluder and the super, so that the two supers and the crown board can be lifted off as one unit. If it isn't too heavy (two full supers will weigh more than 20kg), lift this unit and place on the upturned roof. Don't worry about slotting it into the roof, just place at an angle such that there are eight contact points. If this block is too heavy, then split the joint between the two supers and remove each individually and stack on the roof. Remember, there will be a population of bees in the supers, so be gentle when stacking the boxes.

You will now have the queen excluder exposed. If needed, this may be a good opportunity to puff some smoke across it to subdue the bees a little. Place the hive tool below the queen excluder, at a corner, and break the seal between it and the brood box. You may need to slide the tool around the joint, as the bees often place a lot of propolis in this area. Carefully lift the excluder off, and immediately inspect the underside of it, to ensure the queen is not on it. If she is, encourage her into the hive. The queen excluder should then be inverted and placed on top of the supers. This step is important, as you may have missed seeing the queen on the excluder, and if not inverted she will walk into the supers, then start laying eggs in there – this ruins your cunning plan to efficiently collect honey, and leaves the brood box without fresh eggs/brood. You can now continue with your hive inspection (from the 6th step on p.101).

When your inspection is complete, reassemble the hive by carefully placing the queen excluder, supers, crown board and roof in sequence back onto the hive.

It is not important to look in the supers during every inspection, as we are mostly interested in what is happening in the brood box – so this step can be avoided. However, it is worth having a quick peep at them to understand the quantity of stores, if stores are being capped, and if the comb is being drawn in a tidy fashion. If they are full, or filling rapidly, you will also want to add an extra super. This is all useful information to add to your hive notes.

Hive notes

As hive inspections are pretty intense, most beekeepers have a system to record their observations and to act as a reminder of jobs they need to do. We record our hive notes on an individual log sheet, which we keep in a plastic cover (with a pen) inside each hive, on the crown board. This gives us comparable data for each hive.

The data sheet we currently use is shown opposite. It enables specific details to be recorded for comparison on each visit (14 weekly visits fit on an A4 page), and has space to record other observations and tasks for the next visit. It is completed at each visit to a hive, which is usually weekly from April to September. At the top of the page we record a label/name for the hive, the year of origin of the queen, and the colour we've marked her. The categories we use are described below, but most columns are graded from 1 to 5 to indicate Bad to Excellent for the category:

Queen NS = not seen; S = seen; NL = not laying; L = laying. So, NS/L means we didn't spot the queen, but we did see eggs (indicating she was there in the last 3 days). Whereas S/L means we spotted her and she was laying.

Queen cells Number found. If any are seen, it almost certainly indicates the beekeeper will have to take an action to avoid swarming.

Room Rating from 1–5, denoting available space for hive activities to take place. If all the cells in the hive are filled with brood, nectar and pollen, then it would be marked as 1 (Bad) and frames may need to be manipulated to create space.

Development Rating from 1–5, noting if the hive is growing in strength. Lots of bees and brood would be a 4 or 5, suggesting the hive is increasing appropriately and looks to have an active queen producing eggs, larvae and sealed brood.

Stores Rating from 1–5, indicating the amount of nectar/honey and pollen. Ideally you will see capped and uncapped stores on several frames in the hive.

Docility Rating from 1–5, suggesting whether the bees are calm or aggressive. If they are attacking and/or we are stung on our gloves we'd rate them as 1 or 2.

Brood pattern Rating from 1–5, describing the demographic range of bees. If there is a good number of frames with a nice pattern of brood in all stages it is sometimes denoted BIAS (Bees In All Stages).

Pollen storage Rating from 1–5, indicating amount in cells. The bees need pollen as a protein source, so we expect to see portions of several frames with pollen in plenty of cells. Sometimes it's interesting to note the colour to give an indication of the plant source.

Comb build Rating from 1–5, indicating the comb in situ. If bees are building the comb nicely on the frames with new foundation it will usually also indicate that there is a good nectar flow.

Supers The number of supers that are in place in the hive.

Disease Any signs of disease/parasites. Listed offenders spotted, including chalkbrood (CB), varroa (Var) and/or nosema (Nos).

Weather Temperature and prevailing conditions. This is useful to track because you can sometimes correlate the temperature and weather with bee activity/behaviour. For example, cold weather may cause a low score for docility.

Below these categories, there is space for other observations that aid tracking progress, such as how many seams of bees are present, are the supers receiving any stores, which plants are in flower, any tasks we've completed etc. We also add reminders for jobs to do next time – e.g. add supers, mark the new queen etc.

Each time we visit we scan through the previous comments and scores, and bear them in mind when inspecting to recognise if the colony is improving or not. This is where consistency of your scoring method is most beneficial. We try to keep all our previous record sheets with the hive, enabling comparison with the previous year to take place. This serves as a useful reminder for timings of nectar flows, rate of colony expansion etc.

Typical hive notes sheet used at River Cottage

Hive: Oak tree apiary, hive no.3					Queen: 2021 - marked white					Hive records	
1 = Bad 2 = Unsatisfactory 3 = Average 4 = Good 5 = Excellent											
Date	**Queen**	**Queen Cells**	**Room**	**Develop-ment**	**Stores**	**Docility**	**Brood pattern**	**Pollen Storage**	**Comb build**	**Supers**	**Disease**	**Weather**
	NS/S/NL/L	No.	1-5	1-5	1-5	1-5	1-5	1-5	1-5	No.	CB/Var/Nos	
1/5/23	NS/L	0	4	3	3	2	3	3	2	0	OK	Cloudy 12°C
Fairly aggressive, but quite cool weather. 6 seams of bees, but plenty of brood and stores. Plenty of pollen being brought in. New frames added last week starting being drawn out.												
8/5/23	S/L	0	3	4	4	3	3	4	3	0	OK	Sunny 16°C
Bees calmer than last week. Starting to fill cells with nectar - maybe add supers next week if filling rapidly. 9 seams of bees.												
16/5/23	NS/L	0	2	4	4	4	3	3	4	2	OK	Sunny 18°C
Nectar flow appears to be increasing. Added queen excluder and 2 supers. 10 seams of bees, hive looking strong. Check next week if stores in supers.												

Addressing beekeeping challenges

One of the joys of beekeeping is using your experience, creativity and logic to benefit the apiary when challenging opportunities arise. In time, you will need additional equipment beyond a single hive to address issues, but fortunately, for first-year beekeepers, swarming, or the loss of a queen, are less likely if a newly acquired one-year-old overwintered queen was used to start the colony. For the second year of keeping bees, it's worth acquiring a few pieces of kit.

Swarming

It is completely natural for honeybee colonies to swarm and this occurs from 'wild' colonies as well as those housed in a hive. It has evolved to allow a single colony to reproduce and split into two or more viable colonies. Often it is initiated when the parent colony is living in a space that is too small for the number of bees. This can be due to a physical constraint in the size, leaving little room in the comb for a prolific queen to lay eggs, or simply that a large nectar flow has resulted in the stores collected by the workers filling too much of the comb, with the same effect. Typically, swarming occurs in early and mid-summer (May–July), during the warmest part of the day, and when the colony population is expanding rapidly.

The workers make a collective decision to create a new queen. It will be at least 10 days before she emerges, and up to a further 2 weeks before she has mated and is laying eggs. (If all goes well this colony in the original location will then continue to expand as normal.) In the meantime, and after the first new queen cells have been capped, the old queen and perhaps half of the workers will rapidly leave the hive/colony and head for a temporary place to assemble prior to moving to their new home. Before the workers leave the colony, they will fill their honey stomachs from stores in the hive – ensuring they are carrying enough resources to start building a new home.

This mass of bees, referred to as a swarm, can be quite an impressive sight. Although the swarm looks threatening, and can be regarded as a nuisance, the bees are generally not aggressive, but it is best not to approach or disturb them. They tend to favour settling on a tree branch, or in a hedge – but there are many occasions each year when swarms are photographed at troublesome public locations such as on post boxes, cars, lamp posts etc. If the swarm is accessible, it provides a great opportunity for a beekeeper to collect some free bees (see overleaf).

While in this temporary location, some workers start scouting for a suitable new home, and will then communicate the target place to the swarm through one of their repertoire of bee dances. The swarm will then head there, in the hope that it will be a large and safe enough location. The period until their departure may be a few hours or even a few days.

A small swarm of bees assembled in a hedgerow

Sometimes the new location is problematic for us, such as in a chimney breast, a wall, bird box etc. If the queen has survived this hazardous trip (this may be the first time she has left the hive since her mating flight), there is a good chance this new colony will survive. The workers will rapidly start building comb, using the honey they have stored in their stomachs as fuel. Other workers will start foraging to bring new resources to the colony. As soon as feasible, the queen will commence laying eggs and the first new workers in this home will emerge 21 days later.

There are occasions when several queen cells hatch in the original colony, and one or more of these virgin queens will leave with a significant portion of the remaining worker bees. This is known as a cast swarm, and will be smaller than the original swarm containing the old queen. It will behave in a similar fashion, but egg production will be delayed and it is perhaps at a greater risk of failure as the queen will still need to successfully complete a mating fight.

Collecting swarms

As mentioned earlier, collecting a swarm can be a way for a beekeeper to acquire a new colony of bees to fill a hive. Your local beekeeping association will probably have volunteer swarm catchers. The public can call for advice regarding a swarm they have found locally and/or on their property. If the bees are identified as honeybees (often there are calls regarding bumble bees or wasps), and the swarm is in an accessible location, the swarm collector will likely offer to try to collect and remove the bees – for free!

The procedure involves persuading the swarm to be temporarily housed in a container, which may be a cardboard box, a skep or even a small hive – called a nucleus box. The swarm collector must be wearing the standard protective suit etc., have a smoker running and should ensure that any inquisitive spectators are kept at a suitable distance from the action.

An old sheet can be spread on the ground near the swarm, where the collection box will be placed. The ideal situation is where the swarm is sited up to 2 metres above the ground and hanging on a branch. Next, the collection box is placed, or held, below the swarm and the branch vigorously jolted, such that the whole swarm falls into the box. The lid should be rapidly placed on the box and then placed on the sheet. Sometimes a portion of the branch can be cut off, and the branch and swarm carefully lowered to the box, then knocked so that the swarm falls. A quick brush, using a bee brush, may persuade remaining bees to fall into the box.

The location the swarm came from should be puffed with the smoker, to remove scents from the bees and reduce the likelihood of the bees returning to the same position. Next, the box or skep is inverted, on the sheet, and the contained bees will clamber internally to the top of the box. After a few minutes, the lower corner of the box is then raised slightly to form an entrance hole into it.

If the queen has been captured, then flying bees will pick up the scent of her pheromones and head into the box. Others will position themselves outside (on the sheet) in a radiating pattern, abdomens pointing upwards, and fanning their Nasonov's gland to direct the remaining bees into the container. If the queen didn't arrive in the box, the bees will all leave it and reassemble with her and the whole procedure will need to be repeated.

It is best then to leave the assembly in this position until the evening, when any stragglers will have joined the colony in the container. Once sealed, the bees and container can then be taken away by the beekeeper to their hive location – taking great care that no bees can escape, particularly if placed in a car!

Swarming and queen cells

The first thing to do if you spot queen cells is to assess the situation and see if these cells are capped or not. If capped, then swarming may have already happened, or is imminent. If there is not a queen in the hive, she has probably swarmed or died. An indication of the queen's death may be that no eggs are seen and emergency queen cells have been made (towards the centre of the frames). In this case, you can simply leave the hive for a few weeks for a new queen to emerge, mate and start laying. Some beekeepers recommend knocking down capped queen cells, leaving

A swarm queen cell at the bottom of a brood frame

1 to 3 charged cells where you can see a larva – thus almost guaranteeing they will produce a queen and there won't be a battle between lots of newly emerged queens.

If you can find your swarmed bees nearby, then do your best to collect them – but don't return them to the original hive. When you check out the hive that is missing the swarm, you need to decide if it is likely to be able to survive with the quantity of bees and brood that are left. If it is a weak hive you may wish to merge the remaining bees with another colony (see p.112).

If there are supersedure queen cells (few and towards the bottom of the frames) it may imply overcrowding as the queen cannot lay in sufficient places due to cells being full of brood, nectar and pollen, or the queen has lost her laying ability, is damaged in another way, or she has swarmed. If there are eggs and the queen is found, the best way to avoid swarming is to trick the colony into believing that swarming has already happened.

Splitting with a nucleus (or hive)

There are several ways to manipulate the hive to make the bees think that swarming has taken place. This technique is one of the simplest, but you do need another hive or, ideally, a nucleus box available that is filled with frames of drawn or undrawn comb. Additionally, you need another site available, several miles away, where you can temporarily place this box.

The second site is needed as it will avoid bees taken from the original hive simply flying back to their previous home. You will be splitting the hive into two colonies, one of which will contain the queen, and the other eggs to be used to create a new queen. It doesn't really matter which split has the queen, but traditionally she is removed from her original site. Proceed as follows:

- Place the new hive/nucleus box next to the problematic hive and gently move into it 3 to 6 frames without any queen cells that include nectar and pollen, brood, and empty cells, along with the bees that are on them and the queen. Ensure you leave at least one frame with eggs and one with queen cells in the now queen-less original hive.
- Replace the frames you have removed from the now queen-less hive and ensure both boxes have sufficient bees and resources to survive.
- Make sure the entrance door to the hive that is being removed from the site is closed so that the bees won't escape, and check that the roof is sealed.
- Now move the box/hive several miles away to a new site and leave it to expand.

If all goes well, the queen-less hive left on site will make a new queen. It's best to check after a week or so if queen cells are in place, but otherwise you can leave it for a few weeks for the process to be completed. If you were unable to find the queen, ensure you leave eggs in both new colonies – whichever was queen-less – will create queen cells and subsequently a new queen.

A benefit of this method is that you will have created two colonies – and successfully expanded your beekeeping empire! If you don't want more hives, you can merge the hives back together (as described on p.112) once one queen has been selected and disposed of – ideally to another beekeeper who needs one.

One bit of trickery you may wish to try if the colony that is swarming has an undesirable attribute, such as excessive aggressiveness, is to leave the queen-less hive to develop queen cells, then knock them down after 7 days and add a frame of eggs from a less problematic colony. In the one-week period the eggs and larvae will have either been made into queen cells or matured beyond the age that they can be. Destroying any queen cells stops a queen being created from an offspring of the queen with undesirable genetic traits. When you add the eggs from a 'good' queen, any queen cells made thereafter will be her offspring.

If you wish to return the split hive back to the original site, you can do this after about 4 weeks, once the colony has forgotten about its original location. The hive needs to be sealed during the evening, after all the foraging bees have returned. It can then be returned to the original site, and this colony's bees will reorientate themselves to this apparently new location.

It should be noted that some beekeepers can successfully keep both the hives at the same apiary site, and place the new hive in a different orientation and with some foliage in front of the entrance to confuse the bees regarding their location. If using this method, you should check this hive after a few days to ensure that the population is sufficient and not too many bees have returned to the original hive.

No eggs seen on inspection

If you cannot see any eggs, conduct a thorough hive inspection to check whether there is a queen. Of course, if a queen is not found, she may have temporarily stopped laying but still be there and hiding from you!

If a new queen has emerged from her cell recently but not yet mated, or simply not started laying, the best course of action may be to be patient and leave the colony for another week. However, a useful test to help you out – if you have another hive or friendly beekeeper nearby – is to add a frame of eggs to the hive (without any bees on it). If there is no queen, the colony will quickly use these eggs/young larvae and draw out a queen cell to create a replacement. If there is a queen, they will not make a queen cell, and she may even be inspired to start laying – so check a week later and look for eggs on other frames.

If a queen is found on the inspection but has not recently emerged, and there are no queen cells, leave her for a week and see if she restarts laying. If this does not happen, she will need to be removed and the hive requeened.

Queen cannot be found

It is not always necessary to find the queen when inspecting a hive. If eggs are seen, you know she has been present in the last 3 days and therefore is very likely to still be there. Nevertheless, there are occasions when you do need to find her.

The queen can be very elusive when inspecting a hive as she can move quickly, may be hidden among the other bees, or may be on the brood box side or floor and not on the frame you are inspecting. Of course, it will help if she is marked. In general, she will be found on a frame that is busy, or where eggs have been newly laid. With experience, you will notice her larger size, and that she walks around the frame in a smoother, apparently more controlled, manner.

One useful method to help find the queen is to start by removing 3 or 5 sparsely populated frames from the hive that are queen-less. Then position the remaining frames (8 or 6) in pairs spaced out in the brood box. As the queen will naturally move away from light, she will go into the darker area between one of the pairs of frames. You can then carefully inspect each frame, initially looking on the darker side of it. There will be a greater chance of finding her in this smaller area of comb.

Numerous other techniques exist, some involving using the queen excluder as a bee filter. With a systematic and thorough inspection she will usually be found.

Queen-less hive

If the hive is without a queen and you do not wish to requeen the colony, then there is a way to merge this hive with another and in the process create a single stronger colony. You may wish to do this if a colony is weak, or later in the season when it seems unlikely that a new queen can be created and mated successfully.

Normally, if you were to mix two colonies from different queens they would fight, resulting in lots of dead bees and a weak single colony. The trick is to merge them gradually, so that each set of bees can get used to the other's pheromones/scents. This is done by placing a sheet or two of newspaper between the two brood boxes of bees. It will slowly be chewed through and achieve this slow blending.

Merging hives

This is best carried out in the evening when all the bees are back in their respective hives. Proceed as follows:

- Open the 'queenright' colony and remove the roof and crown board; you may need to smoke the bees down.
- Place a sheet of newspaper on the top of this box, and place a queen excluder over the newspaper to secure it in place. With your hive tool, make a few small slits in the newspaper to allow some scents to pass through.
- Quickly place the queen-less brood box on top of the queen excluder, ideally with its own crown board and roof in place.

'Merged' colonies, which have chewed through the newspaper between the brood boxes

- Leave the hive for a week – look out for pieces of newspaper being carried out.
- Inspect the hive and you will see large chewed-through gaps in the paper and the bees are moving freely through it. Of course, the laying queen will still be present below the queen excluder. You can now remove the added queen excluder and any remaining pieces of paper. Then you can re-order the frames between the two boxes, and potentially even remove the upper brood box.

Weak colony

If a queen can be seen to be laying, but the colony is not expanding in size, there may be a problem with a disease so a thorough inspection is needed. However, early in the season it could simply be that the population of nurse bees and foragers is not high enough to support the number of eggs the queen is trying to lay.

One way to help with this is to add a frame of capped brood from a thriving donor hive. Ideally, you will have a colony in your apiary you can take this from – simply swap it with an unused frame in the weak colony. Once capped, the brood do not require any resources from the colony (other than warmth) and will soon emerge and be able to support the hive activities. There will be no problem with these bees having come from another queen. Just make sure you do not transfer any bees with the frame – they will not react well to mixing with another colony.

Very aggressive bees

Sometimes bee aggression is temporary and relates to changes in the weather, a queen-less hive, or simply a clumsy beekeeper. But, if bees are found to frequently attack the beekeeper/passers-by and/or follow you away from the hive and continue to do so, the behaviour should be regarded as unacceptable. (If bees follow you, it's worth puffing quite a bit of smoke around you, and walking near to, or under, some shrubs to distract these followers.)

Unfortunately, excessive aggression can be a genetic trait of the colony and there may be nothing you can do other than replace the queen to remove this behaviour. This is usually carried out by buying a mated queen from a reputable source, or from another beekeeper who is happy with their colonies. To do this, you need to first find and remove the queen. Almost immediately, the colony behaviour may worsen, and they will be making queen cells. It is time to add the new queen once you have knocked down the queen cells.

Introducing a new queen to the colony

If you bought a queen from a beekeeping supply company, it will most likely be supplied with several attendant bees, in a special introduction cage (called a butler cage) that has one end plugged with fondant. You just need to place this in the

Typical 'introduction cage' containing a newly purchased queen

queen-less colony, according to the supplier's instructions. The colony bees will become accustomed to this new queen's pheromones in the few days it takes for the fondant to be chewed through and the queen released from her cage. She should start laying eggs within a few days of this. However, improvements in the behaviour of the colony may not be seen for several weeks – until her offspring emerge.

If a friendly beekeeping colleague can provide a queen (or you have a spare one), you just need to place her in a butler cage, and use a small elastic band to secure a small single piece of newspaper over the open end. A cocktail stick can be placed through the cage (taking care not to skewer the precious queen!), and used to secure it vertically between two frames in the centre of the hive. The colony bees will slowly chew through the paper and release the queen, and they will accept her presence, having becoming familiar with her odours.

Creating new queen cells

A slower, but acceptable, alternative is to remove and dispose of the queen, then add a frame of eggs from another colony that has better qualities. Queen cells will be created once the queen has been removed. Therefore, after a few days, it is important to knock down any queen cells that are not on the donated frame of eggs, i.e. only retain queen cells originating from the 'nice' donor colony. Then it's simply a matter of waiting 3–4 weeks for the new mated queen to appear. It's worth bearing in mind that the emerged virgin queen will potentially be mating with drones in the same location as the origin of the problematic genetic traits, so there is no guarantee of an improvement.

If you wish to add a frame with a queen cell from another hive in a similar way, it will shorten the time to the emergence of the virgin queen. Care is needed when handling the frame, to ensure the queen larva is not damaged.

Insufficient stores in the hive

At certain times of the year when the availability of forage is limited, or during persistent poor weather, when few bees can fly, it is possible for food stocks in the hive to become so low that the colony will be severely weakened, or even die.

On your hive inspection, it should be clear that the colony has been storing plenty of food and you should expect many of the brood frames to contain capped honey and uncapped nectar. If you notice that these stores are not present and the frames feel quite light, then you need to act quickly, or else the colony may starve.

The simplest solution is to feed the colony with sugar syrup, using a rapid feeder, which is available in a wide range of capacities, holding 2–13 litres of syrup. The feeder is placed on top of the crown board, such that bees can enter the feeder from below and collect sugar syrup from the meniscus in a small ring-shaped area, which avoids zealous bees drowning in a large area of liquid.

To make the syrup, use white granulated sugar and water (not brown sugars or honey, due to the risks from contaminants or spreading diseases). Typically, the concentration used is 1kg sugar to 500ml water. Mix an appropriate volume in this ratio in a large pan and then stir over a medium heat until all the sugar is dissolved; there's no need to boil the mixture. Remove from the heat and allow to cool.

Once cooled, the syrup is ready to use. Position the feeder and add the syrup. Place an eke around the feeder to allow the hive roof to be positioned neatly. It's worth dribbling a little syrup through the centre hole of the feeder into the colony, to enlighten them of the new food supply. When a busy hive is desperate for stores it can empty 2 litres of syrup in a day or so!

Care must taken to avoid spilling syrup near the hive, as this may attract bees from other colonies, or wasps, to rob this new food supply. Don't allow any syrup to be stored in supers, as this will contaminate the honey you collect later. If supers were in use, and are presumably empty, bees should be cleared from them, and the supers removed or placed (sealed) above the feeder and eke. Top up the feeder with syrup repeatedly until a nectar supply is available again, the brood box has plenty of stores, or the bees are not taking any more syrup into the brood box.

The preference for most beekeepers is to leave enough honey stores in a hive, allowing the bees to use their own natural resources – but sometimes this is not possible, and it is better to feed them using refined sugars than to let the colony die.

Stings

It's inevitable that as a beekeeper you will be stung at some time even if you are wearing protective clothing. We are typically stung on our hands as we wear flexible but thin, disposable latex-type gloves, which the bees can easily penetrate. Based on personal experience, it is possible to receive stings to the head, particularly if the bee suit isn't fully zipped up – due to the excitement of a trip to the apiary!

Prevention is the best starting point. It's best to always have a smoker prepared if opening a hive, and to wear a full bee suit with a protective veil. Gloves and wellington boots (with trousers tucked into boots) are also advisable. It is also wise to avoid wearing strongly perfumed deodorants/fragrances or wash your bee suit in scented washing powders, as they appear to be disruptive and attractive to bees. It is also suggested that eating bananas prior to looking in a hive can be problematic, as they contain a compound that has similarities to bee-alarm pheromones.

If there is a visible sting attached to you, remove the stinger and its venom sack from your body as rapidly as possible. To do this, gently but assertively scrape it away sideways with a hive tool, fingernail or credit card. It's best not to pinch it for removal, as squeezing the venom sack will push more venom into you.

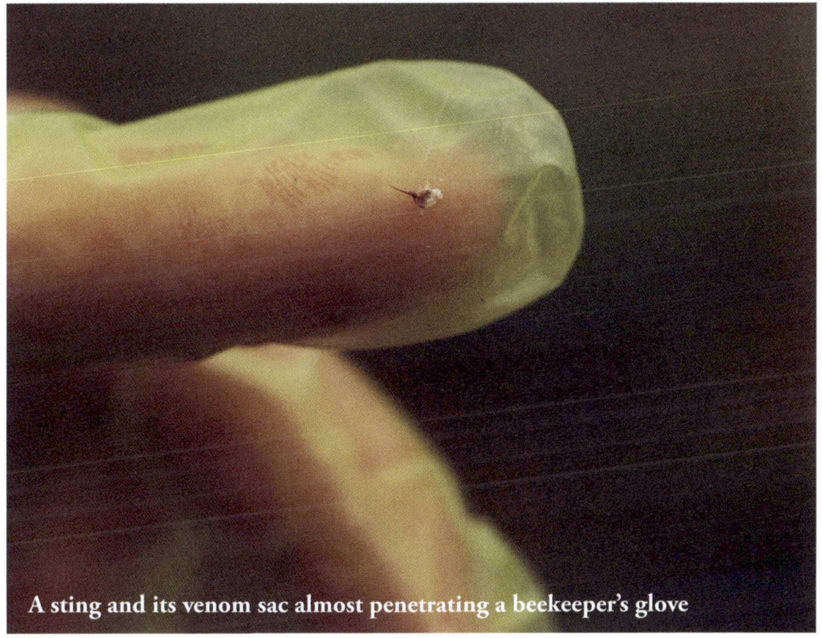
A sting and its venom sac almost penetrating a beekeeper's glove

For the majority of people, a sting is painful initially but the discomfort subsides over 10 minutes or so. Localised reddening of the skin and swelling can last for a day or more. There is some evidence that a level of immunity and weaker reactions occur over time but, unfortunately, you need to receive lots of stings to achieve this!

A relatively small percentage of people (thought to be approximately 2 per cent) will have a significantly more severe reaction, known as anaphylactic shock. Symptoms can include swelling in the throat and upper airways, breathing problems and circulation difficulties, such as tiredness, a fast heartbeat, feeling faint and loss of consciousness. Anaphylaxis is a serious form of allergic reaction and requires immediate treatment – a 999 call to the emergency services should be made, stating your location and that anaphylaxis is the problem. For any beekeeper, it is worthwhile researching the latest recommendations/procedures for sting treatment.

Natural remedies for stings

For a classic home remedy, mix bicarbonate of soda with enough water to make a paste. Apply this directly to the affected area and wash it off after 10 minutes. This remedy is not scientifically proven and eschewed by some because the venom is not on the surface of the skin, but it can calm sting-related inflammation and itchiness.

Diseases and pests

We don't want to give the impression that bee hives are plagued with diseases and pests, but it is important to be able to recognise their presence and effects, and to know how to respond. Some of these problems are commonly seen, treatable and not particularly detrimental to the colony. Others we need to be aware of and looking out for; they are not in the UK yet, but are spreading rapidly across the globe and towards us. A select few – the notifiable diseases and pests – can be regarded as potentially disastrous and may result in the need to destroy the colony and hive to avoid wider transmission.

The following list of diseases and pests is not fully comprehensive but it includes problems that are identifiable by most beekeepers. If you have concerns about a colony's health, you are likely to be able to get advice from your local beekeeping group, or even your regional inspector from the National Bee Unit.

Notifiable or reportable diseases and pests

You must inform the National Bee Unit (NBU) if you identify any of the following:

American foulbrood (AFB)

Spores from this bacterium infect the larvae after worker bees pass contaminated food to them. The infected larvae will die once the cells they are in are sealed, due to the multiplication of bacteria in their guts, which then pass to their tissues. Unfortunately, the spores will remain viable for years, resulting in a simple route for transmission of the bacteria to other hives through robbing by bees, or the beekeeper transferring equipment.

The disease can be identified by an unpleasant smell, patchy brood and the presence of sunken cappings on the cells that are sometimes perforated by the workers chewing into the cell to remove the dead larvae. The simple ropiness test is useful for identification: a matchstick is lowered into the cell, and when raised it will pull a brown sticky mucus-like thread (up to 3cm) with it.

Once American foulbrood has been positively identified by a regional bee inspector (part of the NBU) the bees will be killed and the hives burnt. There are no treatments available in the UK.

European foulbrood (EFB)

This disease is caused by a different bacterium to AFB but is passed to the larvae in a similar fashion. The multiplying bacteria remain in the gut of the larvae, and result in starvation. This will happen prior to the cells being capped. Therefore, there may be patchy brood, the dead larvae may be visible to the beekeeper, have an unpleasant smell and are seen to be brown and sticky – but they cannot be drawn out with the ropiness test used for AFB. If a colony is not severely infected and still strong it may be possible to treat it with an antibiotics, or move the bees into a new hive with fresh and uncontaminated comb, but heavily diseased colonies will need to be destroyed.

Small hive beetle

Not thought to have arrived yet in the UK, but it is feasible these beetles might soon, through importation. They are approximately 6mm long and 4mm wide, oval in shape and dark brown or black. A distinguishing feature is club-shaped antennae. They may be seen on the comb or hiding in the hive, and create larvae with rows of spines and three pairs of legs near the head. The larvae cause extensive damage to the comb by tunnelling, and ruin the honey through contamination with faeces and causing its fermentation.

Tropilaelaps mite

These mites are not currently present in the UK, but their global spread has increased rapidly and they are considered an importation threat. They reproduce in sealed cells, and then spread through a hive and transfer to other cells when the bee emerges. They cause weakening and potential collapse of a colony in a similar way to varroa mites. They are only about 1mm long and smaller than a varroa mite, but can be seen running around on frames.

Varroa mite

These mites are commonly found in hives, to the extent that they are perhaps in the majority of hives in the UK. Despite this, you are still required to report their presence to the NBU. A simple way to do this is take the opportunity to inform them of the location of your hives on 'Beebase', and note it in your records. This also has the benefit of you receiving warning notifications if other notifiable diseases are found in your area, suggesting further vigilance. Varroa can seriously deplete a colony's strength and may be a source for other problems.

Varroa is a parasite that can cause a colony to weaken, or even completely collapse. Present in most colonies, these mites are difficult to control through natural or chemical treatments. The varroa lay eggs within the cells of brood, which then hatch and feed on the capped larvae. Clearly this will hinder the development of the larvae, but can also be a route for viruses to enter the hive.

Varroa are difficult to see in a hive, but can often be found if opening capped cells of drone brood – they are quite visible against white larvae. An individual mite is about 1.4mm in size, oval in shape and has a brownish colouration.

The use of open-mesh floors in hives can help to reduce varroa levels, as those that fall from the colony are likely to pass through the mesh and not climb back in.

Varroa mite on a bee's thorax

In any case, these floors can be useful for indicating the level of varroa in a hive. The insertion of a board that slides below the mesh and collects the fallen mites for potential monitoring is also an effective way to remove them.

Natural varroa deterrents include formic acid, which is produced by ants, which are often seen within hives. Essential oils from lavender, laurel and thyme may also be helpful in fending off varroa infestations.

There are more than a dozen approved varroa treatments in the UK, mostly based on the use of one or more of the following active ingredients: thymol, formic acid, oxalic acid, taufluvinate and amitraz. In the UK, only treatments authorised by the Veterinary Medicines Directorate (VMD) can be used in your hives, and they should be used in accordance with the manufacturer's instructions – ensuring effectiveness and minimising the risks of the development of mite resistance to the specific treatment.

The approved treatments can be found in the VMD website, within the Product Information Database. They are applied in different ways, including placing strips or trays inside the hive, vaporisation of the product, and dripping a solution into the colony. Generally, they should not be applied when there is the potential for an imminent honey crop, so application is typically between late summer and winter, and ideally when supers are not on the hive.

Yellow-legged/Asian hornet

European hornet (for comparison)

Honeybee (for comparison)

Yellow-legged/Asian hornet

Although not currently assigned as a notifiable pest (but perhaps will be), this hornet (*Vespa velutina*) is potentially an enormous threat to many insect species in the UK. It is an aggressive predator of native insects and it's estimated that a single colony can consume 11kg of insects each year. Honeybees are a particular favourite for them to target.

It is important for beekeepers to be able to identify the Yellow-legged hornet. The key features are that they are smaller than the European hornet (*Vespa crabro*), and differ from them by having yellow (rather than black) ends to their legs, as well as a darker body. (It is suggested to download the Yellow-legged Hornet Watch app to help you identify and record hornets.)

Yellow-legged hornets have spread across large areas of Europe, since the arrival of a single queen in Southwest France in 2004. They started to appear in England in 2016 when one nest was found and destroyed. In 2023, this had increased to 72 nests in 56 locations (mostly in Southeast England). Although clearly a rapid expansion, it is important to note that the single colony in France has now expanded beyond to Spain, Italy, Belgium and Germany. In some areas of France, street food markets have been driven indoors due to the number of hornets feeding on the food stalls. Fruit (including grape) harvests are being reduced, and stinging is becoming more common – even disrupting outdoor events. The high number of nests now reported in northern France and Belgium will inevitably result in proportionally more hitchhikers transferring across the Channel to England.

The danger to honeybees is due to their specialism of seeking bee colonies and predating on flying bees. Initially this is from hawking (catching bees on the wing outside the hive), but later in the season from direct attacks into hives. This causes significant stress for the bees – and results in a weakened colony that becomes even more vulnerable to a sustained attack, reducing its ability to cope with existing pathogens. Ultimately, the outcome is insufficient colony strength to survive the harsh winter conditions.

The life cycle of the Yellow-legged hornet is key to its rapid spread. In autumn a colony will release up to 1,000 potential queens (known as sexuals), who will mate with males and then hibernate individually over winter. Although less than 10 per cent of these foundress queens may survive the winter, it only takes one to establish a new colony. In spring the surviving fertile queens will become active and create a small primary nest in a sheltered spot. Workers will start to appear from May, and once the population has expanded enough (July–August), the workers will create a secondary nest that they will move into. This then expands rapidly through the year, creating many workers that try to predate on insects, probably peaking in size in September. These colonies then start their cycle of releasing potential queens.

Strategies to reduce the threat of Yellow-legged hornets are being developed. This includes trapping the foundress queens in their solitary and vulnerable phase in spring and destruction of nests later in the year.

For beekeepers it is important to reduce the attractiveness of an apiary to Yellow-legged hornets by minimising their olfactory visibility (smells) through reduced opening of the hives – when moving honey stores or feeding syrup, for example. A strong and healthy colony will always have a greater chance of resisting an attack. In France and Spain devices have even been successfully developed to electrocute the foraging hornets (known as harps), as well as designs of protective cages around hive entrances to give the bees a greater chance to rapidly exit and enter the hive without being attacked.

Managing the Yellow-legged hornet threat will clearly require a national effort and significant resources to develop technology, trap queens, destroy nests and educate the wider population.

Other diseases and pests

Although the following diseases and pests are non-notifiable, they can significantly affect the wellbeing of a hive, so it's important to be able to identify them and take steps to eradicate them.

A bee suffering from the effects of deformed wing virus

Deformed wing virus (DWV)

This virus results in the death of some larvae and bees having very badly deformed wings. Although there are several ways for it to be transmitted and passed between bees, there is a strong correlation with the parasitic behaviour of varroa mites and their ability to transfer diseases. It is particularly noticeable during large infestations of varroa. Presently, the best way to suppress DWV is to decrease the number of varroa mites in the colony using one of the available treatments.

Sacbrood

The name of this virus is derived from the appearance of larvae that are infected and have died after being sealed in their cells and prior to pupation. The dead larva changes to a brown/black colour within a sealing of its skin. When removed from the cell this appearance is likened to a Chinese slipper. Sacbrood can be spread by varroa mites, so controlling them is beneficial, as well as replacing the queen.

Chalkbrood

This fungal infection causes the larvae to die prior to capping. The dead larvae are heavily infected with the spores that caused their death. Chalkbrood can easily be identified by the beekeeper as the infected larvae will harden in their cells and have a mummified, chalky appearance. Often these bodies will be seen scattered on the hive floor, prior to their removal by the bees. If well established, the sealed brood will have a pepper-pot appearance, i.e. patches of empty cells within frames of capped brood. The disease can be just a background issue, but will be problematic if the colony is weak. Damp conditions appear to worsen the spread of chalkbrood. Requeening the colony is regarded as a useful way of resolving the problem.

Wax moth

There are two moth species whose eggs hatch into larvae that then cause significant damage to the wax as they eat and burrow through it. They will not cause problems for a strong bee colony, but weaker colonies will not be able to overcome the damage. However, beekeepers find wax moth most frustrating when the larvae infest stored comb, as over winter the precious comb can be destroyed.

It is possible to treat comb with acetic acid vapours prior to storing the supers and brood boxes to kill the larvae and eggs, but care is needed with this procedure. It can prove useful to ensure stored equipment is sealed to stop moths entering to lay eggs, and also placing boards between the boxes so that larvae present in one box cannot pass to others. The moths are attracted to the smell of wax, so beekeepers need to be careful not to discard wax near the hives.

Damage to brood comb caused by a wax moth

Nosema

This is a fungal infection, which multiplies in the cells of the gut lining of the bee and causes dysentery. The spores are excreted by the bee, and then easily transferred to other bees. The presence of the disease is typically indicated by the presence of brown spots and lines of faecal matter inside and on the outside of the hive. The colony can be weakened and in severe cases will not expand, or will even fail completely. There is not a recognisable change in appearance of diseased bees, but there is a test that can be used to identify the fungal spores under a microscope.

Mice

Mice can be problematic in the winter months, as they enjoy the warmth and dryness of a hive, along with a buffet of honey and beeswax. It is somewhat surprising that they can squeeze in through an entrance block, and are not attacked by the bees. The best preventative measure to take is to add a mouse guard to the hive entrance in the autumn. This is a perforated metal strip that is pinned over the hive entrance and blocks out the mice, but allows the bees to pass through. The guard can be removed in spring, when the mice are less likely to look for such a warm residence.

Green woodpeckers

Although beautiful, green woodpeckers are potentially a nuisance in some parts of the UK during winter. They feed primarily on ants, but are partial to bees and larvae, mostly in the colder months. They will damage a hive and contents by pecking a hole through the side to gain access. This can be prevented by wrapping chicken wire, or even thick plastic sheeting, around a hive, such that the woodpecker beak is not able to pass through the timber.

Wasps

Towards the end of summer when their colonies are at their strongest, wasps can be a serious threat to bees. They will persistently try to enter a hive, despite the guard bees attacking them. They are attracted to the scent of honey, but will take young larvae from the hive as a protein source for their young. They will weaken a hive, due to multiple bees battling to the death trying to defend the hive, along with the removal of larvae. Once the population of bees is low, it is very easy for the wasps to decimate the colony.

Many beekeepers will use baited wasp traps to lure wasps into a container, where they will eventually die. However, these traps can have a significant bycatch, and kill many innocent flies and moths. An alternative is to reduce the size of the hive entrance to improve the bees' defence ability, maintain a strong hive and avoid any spillage of honey/wax near to the hive.

The Beekeeper's Year

What do beekeepers do? In summary, our objective is to aid the colony to survive healthily through the year, facilitating its expansion to make the most of nectar flows and ensuring it can survive until the following season. Of course, there are many aspects to this that are influenced by the chosen beekeeping philosophy. For example, you may wish to simply encourage the natural seasonality of a colony by providing a home in a Freedom hive or similar, and primarily observe their behaviour without taking any of their honey and with minimal disturbance to the colony.

Alternatively, you may wish to push the bees to be highly productive, to generate a large crop of honey to be sold in bulk. In reality, most beekeepers fall between these extremes, and gain pleasure equally from observing and interacting with the bees, and gathering a smaller crop of honey for themselves. It is this approach we will outline below.

Once we have a created an apiary site and hived a colony or two – what next? Our starting point is recognising the stage in the season and understanding the actual versus expected condition of the colony at this time.

Colony activities through the year

The honeybee colony's objectives across the seasons are as follows:

Spring

- Expanding the number of bees (colony size) to ensure the maximum number of foragers coincides with highest nectar flows, to create plentiful stored food to support an increasing number of bees and larvae. (It takes 6 weeks from an egg being laid to the emerged bee starting to forage.)
- Protecting the queen to allow her to lay eggs without interruption, and replacing her if necessary
- Recognising the need to reproduce via swarming if required
- Producing drones to mate with virgin queens and pass genetic characteristics on to future colonies
- Protection from predators.

Summer

- Highest period of food store collection
- Swarming if needed due to insufficient space in the nest/hive – finding a new home and leaving some of the original colony to create a new colony in the original location
- Protection of the queen

- Drones continue attempts to mate
- Protection from predators.

Autumn

- Reduction in egg laying by the queen as fewer bees are needed, owing to the reduced forage
- Pushing drones out of the hive as they are no longer needed for mating
- Preserving sufficient, accessible stores for the reduced colony size to survive through winter
- Protection from intruders.

Winter

- Survival mode, with dwindling numbers of bees trying to ensure the queen survives the cold of winter until spring
- In late winter, when it begins to get a bit warmer, the queen slowly starts laying eggs again.

Left on their own the bees will follow this sequence without any human intervention. After all, many colonies live in the wild without our help. But if we wish to take their honey (and other products) for our own consumption, we need this to be accessible, hygienic and plentiful. Our beekeeping activities are based around achieving this.

The beekeeper's role

Our act of farming bees gives us a responsibility to respect and care for them – and if successful this will give us the greatest opportunity to harvest some honey. The hive design provides a relatively safe environment to facilitate the desired productivity, but the beekeeper must check all is well with the colony and ensure the home provided is optimised for the purposes intended.

Conventionally, checking the hive consists of a thorough weekly inspection from spring into autumn. We then respond to our observations. For example, during the period of maximum laying by the queen we must confirm there is plenty of space in the hive on empty frames for her to lay, we must check that there are sufficient stores of nectar/honey and pollen for the bees and larvae to survive, and that there is an appropriate quantity of nurse bees available to manage these larvae. This can be achieved by feeding the colony if resources are low, adding frames of drawn comb to aid laying, and even adding bees to the colony if there are insufficient bees (known as merging, see p.112).

If the queen is no longer present, we must allow the colony to produce a new queen undisturbed, or intervene by providing a replacement queen. If swarming occurs, we can collect and rehouse the swarm – or just help the remaining bees to survive. Although tricky, we should aim to avoid swarming altogether by recognising and responding to signs that preparations for it are being made.

It can be argued that the intensity of conventional beekeeping practices makes the colony more vulnerable to failure from diseases etc. The beekeeper needs to be able to recognise these problems and respond in a way that allows the colony to survive. This may well be through chemical treatments at the appropriate time, that are applied responsibly and in a way that does not adversely affect the honey we collect.

The following calendar of activities gives an indication of the activities through the beekeeping year.

The beekeeping cycle

The annual beekeeping cycle echoes the gardening year, of course, and similarly there is variation depending on your location and shifting weather patterns. The last few years have certainly thrown cycles off-kilter.

Winter is a time for rest and preparation ahead of spring, when everything bounces back to life as nectar-rich flowers bloom and tease the hibernating bees out from their hive. Winter is the season to prepare and plan; spring is the time to take action.

If you decide to set up a hive, it's important to carve out time to dedicate to your bees. Once the colony is established and happily thriving, you can develop a more laid-back rhythm to your beekeeping duties. An hour or two once a week throughout the year is the perfect pattern to adopt regardless of the type of hive you have, as even hives housing wild colonies should be checked to ensure your colony is happily buzzing. Spending time with your bees enables you to make sure they're happy and safe.

January

All hives

- Visit your hive a few times throughout the month to ensure the hive is properly sealed and secure from possible bad weather (wind, rain, snow). Strong winds at any time can be a problem so it's best to ensure the hive stands are secure, and the hives strapped to them.
- Check for predators, especially low-to-the-ground hives, particularly for nesting species seeking warmth.

- On a sunny day, check to see if you have any bees out foraging pollen from winter-flowering plants.
- Make sure the water source near the hive is in a sunny spot that will warm the water before the bees drink it, as chilled water can disable bees from flying back to the hive.

National hive or similar

- The bees will consume about 11kg of stores in January so check they have enough by hefting rather than opening the hive as that will bring in too much cold air.
- If the weight of your hive feels light, you'll need to feed your bees. Hopefully you've left them with enough stores to avoid this (as sugar can cause bees gut issues) but if not feed them with fondant rather than let them starve. There are specialist beekeeping fondants, but many beekeepers use baker's fondant, which is sugar with a little glycerine added to soften it.
- Build any hive parts needed for the year ahead, such as new frames, supers or brood box, and repair equipment in storage, such as dry supers.

February

All hives

- Pollen loads being taking into the hive indicate that the queen is laying and brood are being reared inside the hive.
- Make sure that your source of water near the hive is not frozen and is in as warm and sunny a spot as possible to ensure the bees are not drinking chilled water.

National hive or similar

- As winter edges into spring, the queen starts laying and the bees will be getting through more of their winter stores, so check the weight of the hive again to ensure they have enough food stores.

March

All hives

- When it's sunny and 10°C or warmer, visit your hive to see if you have bees venturing in and out – there should be a buzz of activity.
- Also ensure you have flowers for them to forage blossoming around you.

National hive or similar

- If using a National or similar hive, heft the hive to ensure the colony has sufficient stores of food.

- If the bee colony seems light, consider feeding with a sugar syrup made with 2 parts sugar and 1 part sugar using a rapid feeder (see p.89) over the winter cluster.
- If this is your second year with your hive, on a warm sunny day, use a little smoke and lift the inner cover of your hive to peek into see your bees. Look for visible excreta on the top bars, which is a sign of dysentery, i.e. your bees might have nosema (see p.125).
- Don't pull out frames from the brood box until the weather is comfortably above 15°C. It's often said that if it's warm enough to go out in your shirt sleeves it is OK to inspect the bees (but wear a suit!). You also want to ensure there is plenty of blossom (look out for the red flowering currant, (*Ribes sanguinium*), a botanical signal that the honeybee season has begun.

April

All hives

- Plant bee-friendly flowers around Easter weekend.
- Consult an expert from a local beekeeping society if you have any concerns about your bees.

Warré hive

- Add a new brood box to the bottom of the hive to give your colony room to expand.
- If your hive has a window, peek in to ensure the bees have plenty of honey stores and the hive looks healthy.

Top Bar hive

- Expand spacing to let your colony grow, by moving the spacing board a few frames down.
- Check to ensure the bees and the hive conditions are healthy: golden comb, no debris or signs of disease, plenty of honey stores.

National hive or similar

- In mid-April, or when there's a pointed shift into spring (warmer, sunnier, more blossom), check your hive regularly – every 7–9 days.
- Keep hive records.
- Look out for queen cells or other signs your bees might swarm.
- Add a queen excluder and super when the brood box is full of bees.
- Remove any old frames that have old undrawn comb or foundation that is very brown, holey or warped, and replace them with frames with fresh foundation.

May

All hives

- Monitor activity in and around your hive to ensure bees are coming in and out with pollen.
- Check that the bees look healthy and the hives are secure.

Top Bar hive

- Expand spacing as the colony grows.
- As you adjust the spacing, check to ensure the bees and hive conditions are healthy.

National hive or similar

- Inspect your hive every 7–9 days.
- Make sure the colonies have sufficient room.
- Add supers as each new box fills with bees and honey.
- Ensure you have a brood box and frames ready to create an artificial swarm, if needed.

June

All hives

- Make sure you have plenty of blossoming forage for your bees as this can be a lean time for the availability of nectar and pollen: it is often referred to as the June gap, i.e. the hungry gap for bees. This is something to consider well in advance to ensure you've planted enough suitable forage, or situated your hive in a position that is rich with bee-friendly plants right through the year.
- Monitor activity in and around your hive to ensure bees are coming in with pollen, the bees look healthy and the hives are secure.

Top Bar hive

- Expand spacing as the colony grows.
- As you adjust spacing, check hive conditions to ensure bees and conditions are healthy.

National hive or similar

- Inspect your hive every 7–9 days.
- Swarming is still a possibility in June. It's good to ensure there is plenty of space for honey stores in your supers. If too much nectar goes into the brood box, there will be no space for the queen to lay eggs, which can trigger swarming.

July

All hives

- Monitor activity in and around your hive to ensure bees are coming in and out with pollen, the bees look healthy and the hives are secure.

Top Bar, Warré, National hive or similar

- Monitor your honey stores. If you have sufficient stores, this is normally the best month to harvest honey, but make sure the comb is capped.
- Ensure your bees still have space to lay down more honey stores if you've harvested honey.
- Make sure there's still ample nectar, like borage, linden and clover, available for the bees to bring in to the hive.
- Watch out for robbing, i.e. when honey from your hive is stolen by other bees. Note that robber bees never carry pollen into the hive. Other signs include: a frenzy of activity around the hive; robbers looking for cracks and flawed seams to gain entry; fighting bees and dead bees around the hive.
- At this time of year, the brood nest starts to contract.

August

All hives

- This is not a job to do in August but more something to be aware of: some colonies may choose to supersede their queen at this time of year so they can go into winter with a new queen. If the bees have decided that a new queen is required, it is best to leave them alone to complete this process. It is entirely natural and often completely uneventful.

National hive or similar

- As the queen's laying rate reduces, a higher density of varroa is likely because of the reduced number of occupied cells and because the mite is continuing to expand in numbers. So, varroa treatments are advisable (see p.120).

September

All hives

- This month marks the transition from summer to winter as the bees respond to shorter days and cooler nights. There are several important shifts within the hive. Most notably, the colony consolidates, with bee numbers reducing due to less egg-laying and hatching, and drones start getting the boot.
- Your bees may become more defensive this time of year, due to less nectar and pollen being available and more foraging bees in the hive protecting their valuable winter stores from robbing.

- Weak colonies are typically more susceptible to robbing, particularly if the entrance is too large to be defensible.
- The bees tend to store more pollen than usual at this time of the year in anticipation of having to feed the small number of brood raised over winter, but also for general colony nutrition.
- You may notice the bees collecting propolis to fill any small gaps, cracks or crevices within the hive to insulate it and make it more secure in winter. Plugging the draughts and gaps will improve insulation and protect the hive from potential overwintering intruders. Sometimes bees will use propolis to reduce the entrance size themselves, which is an indication that it's a good time for you to assist by adding the entrance block.
- Wasps become quite aggressive in this period and will attack the colony.

National hive or similar

- Consider reducing the size of the hive to give the bees just enough space – no excess will make it easier for them to maintain their warmth over the winter. Many bee colonies can successfully overwinter their cluster and all their food requirements within a single brood box so long as it's full of honey.
- A colony needs 18kg of honey stores to get them through winter – each standard National brood frame holds 2.5kg honey (each 14 x 12 inch frame holds nearly 3.75kg). So, you need eight standard National brood (or five 14 x 12 inch) frames of honey.
- Consider feeding the bees with a 2:1 ratio sugar:water syrup if there are not enough stores. Although it is best if honey stores are utilised, it is preferable the bees survive on sugar rather than die.
- Add an entrance narrower if you are concerned about robbing bees or dropping temperatures.

October

All hives

- The numbers in the hive continue to drop as a smaller worker bee force helps conserve hive resources.
- Winter bees consume less food daily and because they forage and work less, they live much longer than their summer-born counterparts.
- Drone numbers drop further as the weather cools.
- As floral foraging opportunities drop, the bees will venture out less, their flight duration will decrease and their foraging range will become increasingly limited.
- The hive entrance will usually be more heavily guarded to protect the colony from opportunistic intruders and robbers seeking food resources.

National hive or similar

- Ensure you have adequate ventilation to get the moisture out of the hive, otherwise the bees will get cold. If you have solid floors, top ventilation is required. If you have an open-mesh floor, top ventilation is not required and, indeed, it could cause too much draft.
- Consider insulating the hive's roof to prevent it getting cold (when warm air rises it will condense on a cold roof). Some beekeepers have a 5cm layer of roof insulation in place.
- Put your hive at a slight angle, so drips of condensation on the crown board go down the side of the hive rather than drip into the winter cluster.
- Add a mouse guard at the end of the month when the bees are starting to cluster and are not strong enough to defend themselves.
- Cover your hive with a Bee Cosy (purpose-designed insulated, waterproof cover) or chicken wire to protect it from woodpeckers.
- Take varroa counts and treat with an authorised treatment if necessary (see p.120).
- Clean up the comb; this is your last chance to remove old comb. Note which comb is old and dark and buy the frames and foundation ready to replace them in the spring.

November

All hives

- The bees in your hive are almost all winter bees. Many will live up to 5 or 6 months, surviving to make the colony's maiden foraging forays in the spring.
- Some dead bees may appear on the ground outside the entrance during milder days. This is normal and the bees are just doing their housekeeping when the temperature permits.
- Forage is sparse but your bees will venture out on warmer, sunnier days for winter forage, such as mahonia or ivy.
- Make sure hives are secure against the weather.

December

All hives

- December is the sleepy month. Check your hives occasionally to ensure they're secure and dry.
- This is the perfect month to indulge in learning more about bees by reading or attending talks.

National hive or similar

- Check stored brood boxes and supers for signs of rodents or wax moth.

Adding varroa treatment to a brood box

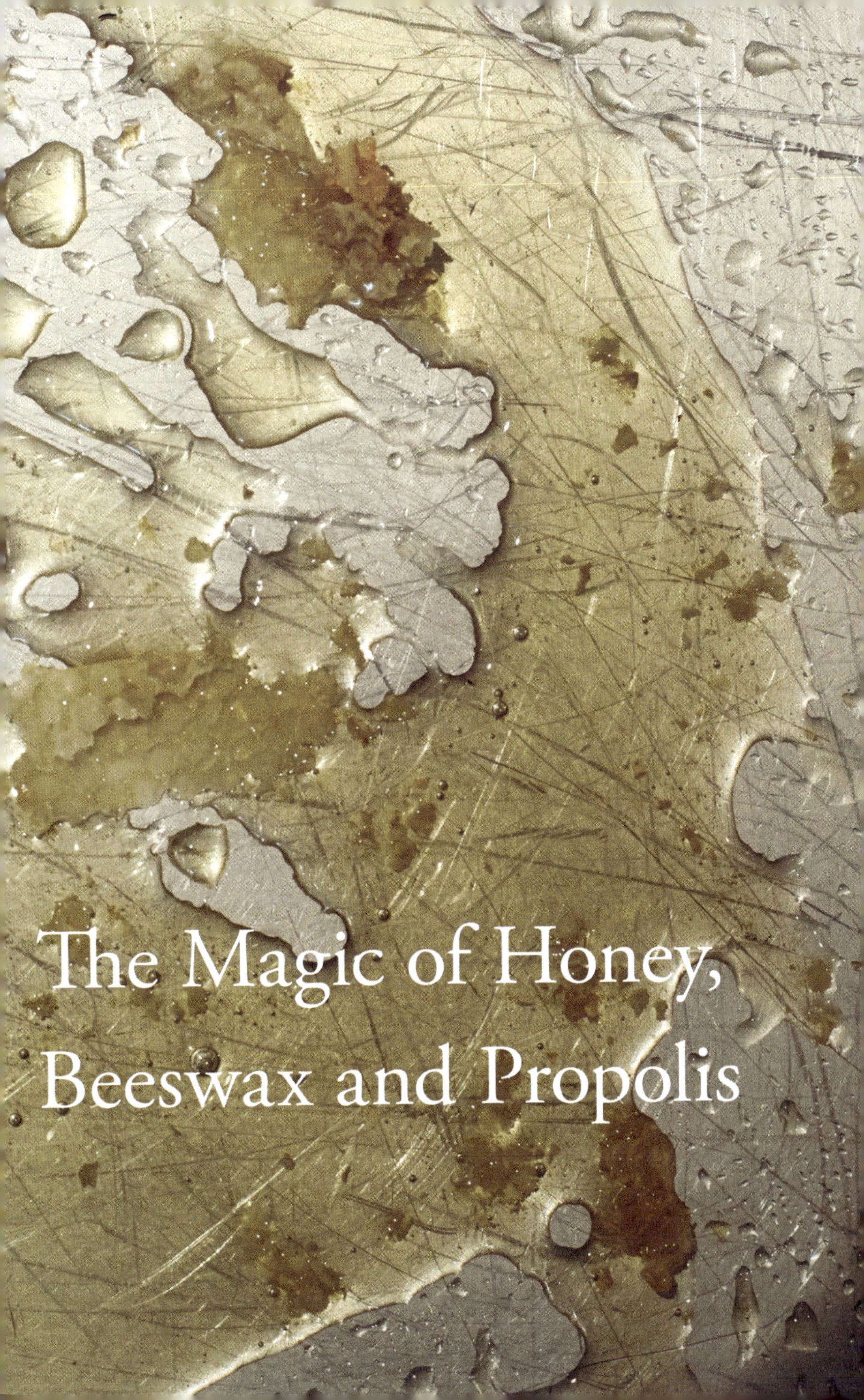

The Magic of Honey, Beeswax and Propolis

Honeybees are pure alchemists, gathering sunlight in the form of golden sweet nectar, pollen dusted on the stamens of flowering plants and dewy sap from trees – to create some of the most nourishing food and medicine that exist in the world, with an eternal shelf life.

To make just one drop of honey, which is enough to fill a cell of honeycomb, it takes around 1 hour and the work of several bees. Beyond honey, the wax created to store the transformed nectar in is also rich with health benefits. It has inflammatory-modulating effects and is used as the base for many herbal salves and topical healing medicines.

Propolis is yet another brilliant product of honeybee ingenuity, made from plant resin gathered from the bark by the bees. It is a vital ingredient in the wax cell walls of the bee hive, preventing them from becoming too brittle and falling apart; bees also use it to seal gaps in the hive. Propolis is also regarded as the quintessential bee medicine: it stimulates the immune system, helping the body to fight off infections and its antiseptic properties promote wound healing.

How bees make honey

When worker bees are about 3 weeks old, most inherit the role of forager (some become guards, but most go out and forage). They leave the hive, typically flying up to 3 miles from their hive in search of the sweet nectar produced by the glands of flowers. Foraging bees leave the hive around 15 times per day and visit up to 10,000 flowers in a single day.

Once a bee lands on a flower, it dips its tongue into the base of the flower to drink up the nectar. As the nectar journeys through the bee's body, water is drawn out before the nectar is deposited in the bee's honey sac or abdomen (also known as a honey tummy). An enzyme called invertase is added to the nectar while it is in the bee's honey sac. When a worker bee has a full load, she flies back to the hive.

While the key component of honey is nectar, it isn't quite honey until the bees add their own magic touch. This alchemy involves the enzymes from their saliva. When a foraging worker bee returns to the hive, she passes the nectar from her full honey stomach, via her mouth, to other worker bees. To the observer, it looks like the bees are kissing. Each recipient worker bee will chew on the nectar for about 30 minutes, mixing the nectar with its saliva, before passing it on to another bee. As the nectar is passed from bee to bee, it is transformed into honey in this way, but it still has a few more steps.

The bees regurgitate the nectar from their mouths and deposit it into the combs – the hexagonal wax cells, which are like tiny jars made of wax. Bees have an enzyme in their stomachs called glucose oxidase. As this enzyme mixes with the nectar and breaks it down, two by-products are created: gluconic acid and hydrogen peroxide. The latter is just one of the substances that fortifies honey, giving it antimicrobial properties.

At this stage, the honey is still quite watery, not the consistency we are familiar with. It's the task of newly hatched worker bees to drive down the moisture content, by vigorously fanning their wings over the nectar, causing the liquid inside the cells to slowly evaporate. The fanning stage lowers the water content from around 80 per cent to 18 per cent. This gives honey its robust shelf life and is essential to prevent it from fermenting into mead, which is toxic for bees.

The drying process, along with the natural enzymes added to the nectar during the chewing process, metamorphose the relatively thin gathered nectar into the thick amber syrup we know and love. Once the honey is ready, the workers cap the cells with beeswax to seal and protect this precious substance to ensure they have food stores for winter, and other times when they need food and don't have access to fresh nectar and pollen.

How much honeybees produce

In a single year, a honeybee colony gathers around 18kg of pollen and 120kg of nectar. Once the nectar is concentrated down to honey form, the honey produced by one colony can be as much as 27kg in a good flowering season but it's important to consider that bees need at least 13kg to see them through the winter. If you accumulate the flying miles it takes a collective of bees to make just one jar of honey, it equates to around 650 bees flying up to 32,000 miles (more than one trip around the world) and foraging from more than 1 million flowers.

Medicinal properties of honey

The complex and considered process involved in creating each drop of honey imbues it with layers of nutritional benefits and healing properties in its purest form. Just one of honey's perks is that it is antiseptic. We've all heard the advice to drink hot water with honey and lemon for a sore throat or cough and there are many studies to back this up. Honey can fight a number of strains of bacteria and it also helps draw water out of tissue, which can reduce swelling and discomfort, if you have a sore throat for example.

Enriching honey's effectiveness are its levels of vitamins B, C, D, E and K, minerals, enzymes and essential oils. Honey is also rich in antioxidants that help it fight respiratory infections and it is brilliant for healing wounds. Driving the moisture out of the honey, as the bees do, fully preserves it and makes it so thick that it rejects any kind of growth. It's also naturally extremely acidic, with a pH between 3 and 4.5, which is a level that will kill off almost anything that wants to grow there. This, paired with honey's traces of hydrogen peroxide, make it an ideal antiseptic – providing a barrier against infection with healing properties. Honey applied to a wound will draw out water, which might be a source of infection, while at the same time releasing hydrogen peroxide, which promotes healing.

The ancient Egyptians used medicinal honey regularly, making ointments to treat skin and eye diseases. They would use it to cover a wound, burn or slash, because nothing could grow on it – so it was a natural bandage. The earliest recorded use of honey for medicinal purposes is found on Sumerian clay tablets dating back to the twenty-first century BC which state that honey was used in 30 per cent of prescriptions.

Honey in its raw form is also employed to attain a glowing complexion and keep the skin healthy. Its potent antimicrobial properties make it useful as a natural antibiotic on the skin and it's also applied to dry skin conditions like eczema.

Ayurvedic practitioners call honey *madhu*, meaning 'perfection of sweet'. They often keep bees in locations with particular herbs to create honeys with unique healing effects. It is believed that certain herbs and flowers have an effect on the qualities of the pollen collected by the bees, giving the honey harvested in the area specific properties. Manuka honey is a great example. It is produced from the nectar of the evergreen, tree-like manuka shrub in New Zealand, which has a higher concentration of antibacterial properties than most plants, and these all go in to the honey.

Honey vs sugar

Honey is man's first true taste of sweetness beyond fruit. Its thick, syrupy intensity makes it a truly addictive substance, but is it good for us and how does it stack up as a substitute for sugar?

Both honey and sugar are made up of a combination of glucose and fructose. In sugar, glucose and fructose are bound together to form sucrose, which comes from sugar beets or sugar cane and is more commonly known as table sugar. In honey, fructose and glucose are primarily independent of each other. In addition, around 25 different oligosaccharides have been detected in the composition of honey. Oligosaccharides are a group of carbohydrates that play an important role in nutrition because they are prebiotic, which means they feed the good bacteria in our gut. (Sunflower seeds and Jerusalem artichoke are other foods that contain high levels of oligosaccharides.)

Honey also has trace elements, gathered by bees while going from plant to plant. These will depend on the actual location, so depending on the source of your honey it could have varying small amounts of minerals like zinc and selenium, as well as traces of B vitamins like riboflavin, niacin, folic acid, pantothenic acid, vitamin B6 and vitamin C. Honey also contains small amounts of proteins and antioxidants called flavonoids, which are known to reduce inflammation in the body. Sugar, on the other hand, has an inflammatory effect on our systems.

The more complex a food is, the more work it takes to break it down, which also tips the scales in honey's favour. Sugar is comprised of 50 per cent glucose and 50 per cent fructose and is broken down very easily in the body, leading to a surge of blood glucose. What your body doesn't use right away gets stored as fat. So even though honey is also made mostly of sugar, it's only about 30 per cent glucose and less than 40 per cent fructose, with about 20 other sugars in the mix, many of which are much more complex, plus dextrin, a type of starchy fibre. This means that your body expends more energy to break it all down to glucose. Therefore, you end up accumulating fewer calories from it than if you were to consume the same amount of sugar.

So, when it comes to getting your sweet fix, honey comes out on top, but it still should be consumed in moderation. The good thing is it tastes sweeter than sugar so you can use less while getting more (nutrients and flavour). The key to fully reap the benefits, however, is that the honey must be in its purest, rawest form.

Pollen in honey

Honey contains very little pollen. If you examine comb in a hive you'll see some cells are packed with the pollen boulders the bees make by mixing gathered pollen with nectar and their own saliva. This is packed down further to make bee bread. The other cells will either contain eggs or larvae, or honey. In such a small space, there's a lot of intermingling.

Cells that contained pollen or bee bread can be cleaned and reused to store honey and this could result in trace pollen in the honey. The bees will also have pollen on their bodies that can get mixed in with the honey. The amount of pollen in honey varies hugely, but analysis of honey composition typically shows less than 0.7 per cent amino acids in the honey, some of which will be proteins from pollen residues.

Unfiltered honey will contain more pollen. Being able to detect some pollen in the honey is significant when testing honey of unknown origin – it's on par with DNA linking the honey with flower sources. The art of detecting and analysing pollen in honey to work out its origin is called melissopalynology.

Honey laundering

Local honey might cost more than the jars of honey in the supermarket, but most of the time you get what you pay for. A large percentage of commercially available honey is aggressively processed. It is heated to such high temperatures and ultra-filtered through miles of stainless steel and then blast-microfiltered to remove the pollens, a tiny but crucial component of real honey. The pollen is filtered out both to delay the natural process of crystallisation and sometimes to remove the pollen evidence which would lead analysts to trace the countries of origin. Often honey from one country is re-labelled as the product of another.

Sometimes commercial honey is also fully pasteurised, eliminating its fragrance and changing the chemical composition of the honey itself, so much so that it loses not only flavour but nutritional value. Powerful antioxidants, enzymes and vitamins are destroyed when heat is applied to raw honey. The antiviral, antifungal and antibacterial properties of intensely heated honey are also lost, which put it more on a par with eating refined sugar.

A lot of commercial honey is also blended, sometimes by mixing honey from several different countries to achieve a standard colour and texture, but also to drive down costs. (This happens with olive oil and wine, too.) Traces of ultra-cheap sugar syrups have also been found in commercial honeys. A European Commission report, analysing honey sold in the EU from 2021 to 2022, found that 70 of the 123 companies assessed exported honey suspected of containing sugar syrups, which can be made more cheaply than the genuine article.

The global flood of cheap, processed honey undermines beekeeping to such an extent that up to 10 million hives may be lost from the EU's agricultural food production within the foreseeable future, according to a report by European farming unions. It's not exaggerating to say that the honey blending industry directly threatens our national food security.

So, if you can opt for local honey, your taste-buds and your body will thank you… as will your local beekeeper.

Raw honey

As a rule, trading standards in the UK do not permit the use of the word 'raw' on a label if the honey is filtered in any way. Most small-scale, local beekeepers filter their honey through a standard kitchen-style sieve, which ensures no wax, dead bees, flies or other debris from inside the hive are present in the honey. This filtering does remove some of the pollen from the honey, but not in the same way that large-scale processing does.

Most small-scale beekeepers don't directly heat their honey. They do often bring the temperature up in the room where they are spinning or pressing the honey from comb, as it helps ensure the honey flows from the wax cells once the wax capping from the top is removed. This is a bone of contention with small-scale beekeepers as it means their honey is not easily distinguished from the more industrial honey processing as described above (high heat, ultra-filtered and sometimes pasteurised).

That said, the purest honey comes directly from the comb and as it's not filtered, it can be labelled as raw, but the second-best bet is to source local honey from a beekeeper you know, and ask them how they extract their honey.

Organic honey

To be organic, apiaries have to be sited on certified organic land and distant from sources of contamination (conventional land, road, industry, etc.), which means it's only really possible to ensure bees are protected from chemical residues on larger organic holdings, or organic sites in very wild areas.

Bees travel up to 3 miles in any direction from their hive when collecting nectar, which means they can cover up to 50 square miles surrounding their base. In order to be certified organic, all the land in this radial area must be fully organic. Sadly, this disqualifies all British bee hives from organic certification as we don't have a density of untreated land to fall within the remit.

Most organic honey comes from huge organic farms in countries like Argentina which have hundreds of acres of organically managed land. A dream would be to make it possible for at least one British county to convert all its farmland to organic so we can then have an organic British honey.

Pesticide residues do show up in honey. After the ban on neonicotinoid pesticide seed dressings, the EU implemented a moratorium to assess the effectiveness of the ban by collecting 130 samples of honey from beekeepers across the UK before the ban (samples were taken in 2014) and after the ban was in effect (2015). Neonicotinoids were present in about half of the honey samples taken before the ban and they were present in over a fifth of honey samples after the ban, which shows that pesticide residues are still showing up as the chemicals linger in the environment for decades.

When you buy honey of unknown country of origin, the risk of pesticide residues in the honey is greater, which is a reason, in my view, to buy small-scale, local honey, and to support organic farming.

From top, clockwise: Avocado honey, Coriander honey, Ling heather honey, Sunflower honey, Cotton blossom honey, Buckwheat honey

Flowers and flavours

There are more than 300 varieties of flowers and blooms that honeybees can visit, which means the flavour, colour and smell will vary greatly from honey to honey. You can easily view honey like wine. As wine has *terroir* (a French term used to describe the environmental factors that affect a crop's character traits), honey is also a delicious consumable capsule of time and place.

The highly differentiated colours, textures and flavours in honey relate directly to the land and soil from which they come, the plants the bees have visited, the prevailing climate and the specific weather conditions of the time. The bees venture up to 3 miles from their hive, so their honey will be a perfect snapshot of that place and time. Sometimes the flavours from a single flower are captured. This is the result of two conditions. First, the target plant must predominate so the bees have little choice of plants. Second, the beekeeper must time the introduction of the hive and the harvesting of the comb to coincide with this blooming period. This is done by carefully observing the blooming period of the chosen plant, as well as possible overlapping blooming periods of other nectar-producing plants.

Monofloral honey is rare, sometimes difficult to achieve and often it's not great for bees. Think of monocultures and the effect that lack of foraging diversity can have on a bee's health. Feasting from a single flower narrows the opportunity for nutrient intake. Blueberries, for instance, are a brilliant source of nutrients, but they will not provide you with everything you need for optimum health. The recent '30 plants in a week' dictate coined by epidemiologist Tim Spector, at King's College London, highlights our need for variety. Bees need this, too.

Manuka honey is contentious in this respect, and this is also where potential honey laundering is at play. According to New Zealand's Unique Manuka Factor Honey Association (UMF), 1,800 tonnes a year of manuka honey are consumed in the UK each year, out of an estimated 10,000 tonnes globally. Yet production of the genuine stuff is set at just 1,700 tonnes, or the equivalent to more than three million small jars. Unless Britain has somehow managed to secure all of it, there's a lot of smoke and mirrors. The real stuff contains an active ingredient called methylglyoxal (MGO), which is often noted on the label. MGO is known to kill bad bacteria on the skin or in the gut, which is why manuka honey is prized for wound healing and improving gut health. Manuka can either be classified as Multifloral Manuka (where manuka is the main floral source above all others) or Monofloral Manuka (where the level of manuka in the honey is higher).

In the UK, honey from a single floral source (aka monofloral honey) is a challenge. This is partly because we are such a small, crowded island and do not have the large areas of wilderness which produce monofloral honeys in other areas of the world. One exception is heather, which grows in remote pockets.

Some bees are moved to heather-rich moors in the summer, to feast on the sun-kissed lush purple flowers during the summer and autumn. This tradition of moving hives to heather dates back many hundreds of years, resulting in what is often described as the king of honeys. Taking hives to the heather requires specialist understanding and management of the bees, and is best left to the experts.

Heather honey is a dark honey with a distinctive full-bodied taste and a thick jelly-like consistency. It becomes less viscous as soon as you agitate it by dipping your spoon in – a property called thixotropic. It is also valued for its health benefits: high levels of antioxidants and antibacterial properties, and it's rich in manganese.

Floral honeys

Sometimes you can access small batches of honey harvested after a proliferation of a certain flowering plant. These rare jars of honey, as with wine, can elevate your honey-tasting knowledge. In Italy, they train honey sommeliers. Sarah Wyndham Lewis of Bermondsey Street Bees is one of them and opened our eyes to a world of exciting honey flavours, colours and textures. The list below might also inspire planting in your garden, to entice your bees to produce some of these flavours.

Light honeys

These are generally milder in flavour than darker honeys:

Apple blossom

Light amber colour; good flavour, often with a hint of apple; granulates quickly. But it's rare to find an apple blossom specific honey.

Borage

Pale, creamy colour; subtle aroma and a light honeydew flavour, with hints of orange and lemon.

Clover

Faint golden hue; soft, gentle floral flavour.

Cotton blossom

Thick and butterscotch-like with an uncanny fluffy, cotton candy-like texture that's soft, stretchy and fibrous.

Oilseed rape

White, pale and thick; soft, subtle, candy floss-like flavour.

Rata

The native New Zealand rata tree only flowers every 2 years, making this a really unique honey – considered one of the finest in the world. Pale colour; subtle, distinctive, buttery flavour with floral notes and an almost salty taste.

Sunflower

Thick and yellow honey, resembling lemon curd – like creamy sunshine in a jar; light, fresh and citrusy flavour.

Medium-coloured honeys

More robustly flavoured than light honeys, but not as strong as dark honeys:

Dandelion

Egg-yolk rich yellow in colour like homemade lemon curd, turning amber as it matures. It has a buttery texture, an aroma of dandelion and a pungent flavour – sweet, sunny and pineapple-y.

Hawthorn

A gorgeous deep amber honey, clear and with good viscosity, with distinctive fruity top notes, caramel undertones and slight hints of citrus.

Orange blossom

Light amber in colour; distinctive sweet flavour and mild aroma, with echoes of orange blossom.

Rosemary

Amber coloured honey, with a pleasing fragrance and a light floral taste that has a distinct hint of rosemary. The texture is relatively thick.

Thyme

Bright amber honey, intensely aromatic with a rich flavour of thyme, herbs and pepper.

Dark honeys

These are usually stronger in flavour than light honeys and richer in antioxidants:

Arbutus

This late-blossoming tree, also known as a 'strawberry tree', produces a strawberry-like fruit. The deep mahogany honey derived from its flowers is exquisite, with a delicious bitter-sweet flavour. Rich in antioxidants, it has therapeutic properties.

Avocado

Dark amber honey with a silky texture and a very bold flavour profile – vegetal with a somewhat burnt, smoky, molasses taste. Initially sweet and intense on the tongue, it releases light bitter-sweet notes deep on the palate.

Buckwheat

Buckwheat produces small white flowers, which bees love to forage on. The honey it produces is strongly aromatic, dark and malty, with toffee notes and a hint of sour rhubarb (which is interesting as it is from the same family).

Chestnut

A rich, dark forest honey, with a strong aroma; distinctive yet subtle flavour, with slightly bitter, burnt-toffee-like notes, as it is rich in pollen and tannin.

Coriander

Dark amber in colour; thick, treacle-y consistency; pleasing floral aroma and a complex yet mild flavour with liquorice notes, not at all reminiscent of the herb.

Heather, ling

A strong-tasting, dark honey, reddish orange to dark amber in colour with an unusual thick, jelly-like consistency. Unique flavour and distinctive aroma, with beautiful aromatic floral notes.

Manuka

An expensive, rare honey, as manuka blossoms are only found in the forests and hills of New Zealand. Orange-brown in colour, it is a thick, rich honey with a creamy, smooth feel in the mouth. Pleasingly aromatic, it has a slightly earthy flavour with soothing eucalyptus tones and a barley sugar finish.

Oak

An indulgent, very dark honey harvested from the honeydew in the bark of ancient oak trees and wild local forest herbs. Amazing aroma and a woody flavour, reminiscent of molasses, vanilla, walnuts and dark chocolate. Thick, grainy texture, so as a raw product it may be difficult to spread.

Sidr

One of the finest honeys in the world, referred to as 'liquid gold' in the Middle East and derived from the nectar of the endangered Sidr trees which grow in the region. Light, golden colour which darkens over time; stunning creamy, quince-like flavour.

Honey classifications

Beyond the predominant floral notes in a honey, there are various other layers to production that affect both flavour and health benefits. If you're looking for the purest honey in both respects, raw is the way to go.

The honey we eat is from broodless frames, i.e. their extra stores of honey. Unfiltered honey taken from brood-free comb gives you maximum pollen content, which ups the good-for-you ante. Opting for local, of course, not only gives you a taste of your own landscape but it can also help you acclimatise to local pollen – in a gentle, delicious way if you suffer from hay fever.

Here are a few other classifications you might see on a honey jar label:

Chunk honey/cut comb in honey This contains at least one piece of comb honey.

Comb honey This honey has been stored by bees in the cells of freshly built broodless combs or thin comb foundation sheets made solely of beeswax and is sold in sealed whole combs or sections of such combs.

Drained honey This is collected by draining de-capped broodless combs.

Extracted honey Obtained by centrifuging de-capped broodless combs.

Filtered honey This describes honey obtained by removing foreign inorganic or organic matters in such a way as to result in the significant removal of pollen.

Honeydew honey This is obtained mainly from excretions of plant-sucking insects on plants, or secretions of living parts of plants. It is inclined to be darker in colour and stronger in flavour, with a more subtle sweetness. Its distinctive aroma can be resinous, piney, herbal or malty, as it tends to come from sap drawn from pines, firs or beech trees. It's more common in Greece and New Zealand.

Pressed honey This defines honey obtained by pressing broodless combs with or without the application of moderate heat, not exceeding 45°C.

Organic honey This description indicates that the bees which have produced the honey only forage in certified organic land. This is tricky in so many places, including the UK, where you can't guarantee the bees won't venture onto flowers on non-organic certified land. There has to be swathes of organic fields far beyond the bees' reach in order for honey to obtain the organic label.

Harvesting your own honey

The traditional time to harvest honey is when summer ebbs into autumn, once the bees have been feasting themselves silly on the nectar of sun-drenched flowers and the hive is overflowing with honey, but there is still a flow of nectar coming in. This said, there's also a practice of waiting until spring arrives to harvest the honey, which means you leave all the stores in place over winter and only take the surplus in the spring once the bees are out foraging and bringing new nectar into the hive. This is a time of the year when the bees are less protective, too. However, most honeys will crystallise over the winter and become much more difficult to extract.

Another time honey is collected earlier in the year, in small batches, is once the blossom of a particular flower fades, allowing one to capture a honey reflective of a specific flavour, colour and textural notes. But this can mean taking honey at a time when there's uncertainty around future nectar flow.

One thing to ascertain before harvesting honey is whether or not the honey is ripe. By that we mean fully processed by the bees (i.e. all the moisture driven off) and capped so it's in a long-life, storable state. There's a simple visual cue to help you detect this: capped honey is white on the surface, whereas comb with a biscuity golden cap is housing baby bees, which you definitely don't want to disturb.

In an average hive on a good year, the bees will produce around 11kg honey surplus to their needs – that's roughly 32 jars (340g capacity). You want to make sure you leave them with at least 13kg of honey to keep them fed over winter. Even if you're harvesting in spring, it's worth ensuring they have some stores as the weather could take a turn (Easter snow is not uncommon) and you need to ensure the bees have sufficient food.

The hive and approach we currently love the most is the Warré hive, where you wait 3 years before taking your first harvest and then you only take one box of honey from the top. This means less honey and a longer wait but, for me, it makes the honey taste all the sweeter.

Before harvesting or eating honey, we always like to take a moment to reflect on the impressive, laborious process. It changes the way you engage with honey. We really do see it as liquid gold. Each cell of honey is roughly one-twelfth of a teaspoon and that equates to one bee's life's work. The nectar in that tiny cell is the result of flights to thousands of flowers. The vast number of bees and work required to create the honey you're about to harvest is staggering.

Other things to harvest from the hive

As well as honey, you can also consider re-purposing some of the bee's wax from the hive, especially if you're using comb from a hive without set frames. The wax is great for making candles (see p.243), plastic-free food wraps (see p.241) and

Uncapping a frame of honey

various beauty products (p.236–9). And there's the medicinal propolis, which you can turn into healing tinctures (p.244) and a lovely infused oil (see p.247).

If you're using comb from hives with frames and foundation, you can return it to the hive. The bees will remove any excess honey, clean the wax and reuse it next year – which gives them a head start in the spring as it means they won't have to spend extra energy producing wax to build comb. Bees need the equivalent of six jars of honey to produce enough wax to make a candle the size of a honey jar.

Timing is everything

Whether you harvest in late spring (taking frames of honey the bees didn't consume over winter) or if you're opting for a more traditional late summer or early autumn harvest, one thing you want to try to factor in to your timing is nectar flow. When the flowers are still providing fresh nectar and the bees are busy bringing new stores in, the bees will be more laid back about you coming into to dip into their stores; i.e. they'll be less protective or, shall we say, less aggressive. Let's face it, if someone just entered our house and started randomly taking things from the fridge, we would be a bit annoyed!

Another important factor to consider is time of day. Early morning or late evening is best as you want to avoid the bees being out foraging and flying about or they'll be after the honey.

Honey extraction

This is a messy job. To maintain domestic harmony, care is needed when deciding the location for the work, and when making appropriate preparations prior to starting. Kitchens can be an ideal location – but be warned that honey and wax will inevitably drip onto the floor and be found on any surface that is touched. Check your shoes to avoid treading it throughout the house!

Preparations

Hygiene is key, as with the preparation of any foodstuff, and even for a small-scale beekeeper there is legislation in place to follow and should be checked via the UK Food Standards Agency. The location should be clean, to ensure there is no chance of contamination of the honey at any point. Lots of clean uncluttered work surface is needed to aid ergonomics and allow a good work flow. Pets and stray humans should be excluded!

It is also important for the room to be sealed such that bees and wasps cannot enter when they smell the honey (they will), so close any windows. A separate water source is needed to clean the floor, and this should be repeated wherever honey or

Inserting an uncapped frame of honey into a radial extractor

wax is dripped. You should wear an apron, and an appropriate head covering to stop stray hairs falling into the product. Ideally, the room temperature should be between 20 and 30°C. If too hot, the wax in the supers can deform too easily; if too cold, the honey will not flow out easily.

Obviously, all of the equipment should be thoroughly cleaned before and after use to prevent contamination of the honey.

Equipment needed

- **Extractor** to take the honey from the comb. There are several options available (see below).
- **Uncapping knife** or similar gadget that opens the capping on the frames. A bread knife works, but cheap designs also exist that are tailored for this use. Electrically heated ones can be also bought.
- **A source of clean, hot water** to wash extraction tools and to aid mopping up inevitable spills.
- **Clean cloths**
- **Large tray** for use when uncapping the frames. This can be a large stainless steel oven tray, or one specifically designed for the purpose with filtration of the wax and honey built in.
- **Stainless steel strainers** in two sizes, which are used together to filter wax and debris from the extracted honey as it flows from the extractor to a collection bucket. Typically, the top filter has 1.5mm holes and the lower filter 0.5mm holes. They are placed together on top of the receiving bucket.
- **Plastic food-safe buckets** to collect the honey. You will find it convenient to initially use a larger one (30kg capacity) with a gate valve fitted to collect the filtered honey. Then you will need several 10kg (or more) buckets to collect the rest of the honey. The number required, of course, depends on the quantity of supers being extracted.
- **A calibrated refractometer** to measure the water content of the honey.

Extractor options

The majority of beekeepers use a rotary extractor to take the honey from the comb. These come in quite a few sizes, denoted by the number of frames that can fit into them. The choice of extractor size mostly depends on the number of hives and their productivity, combined with budget and time constraints.

The cheapest extractors are made of food-safe plastic, can hold 2–4 frames and are manually rotated. But often the most sensible initial purchase is a more robust stainless steel, 3 or 4 frame, manual extractor.

However, for the first year or two of beekeeping, if you're a member of the local BBKA group, it's usually possible to rent an extractor for a small fee. This enables you to gain experience with the equipment without the investment, and allows a considered choice to be made for a later purchase.

There are two different configurations for extractors: tangential and radial. Both have an internal cage that holds the frames, which is rotated at speed to force the honey from the comb. If only a few hives are owned, a manually rotated extractor is fine, but if a beekeeper has six or more productive hives then investing in a substantially more expensive electric version may be worthwhile – these often have programmable settings allowing automatic rotation at several speeds and directions.

With a radial extractor, the cage locates the frames such that they are arranged like the spokes of a bicycle wheel, i.e. they are aligned with a radius of the circular extractor (when looking from above). When spinning, both sides of the frame will lose honey simultaneously, but it takes quite a bit of time for the honey to leave the cells in the comb, as the cells are not aligned ideally for the forces to spin it out.

In the case of a tangential extractor, the frame is positioned tangentially in relation to the drum. This has the benefit of sending the honey directly out of the frame, but the extractor needs to be stopped, and the frame reversed, to extract from the other side. This reversal needs to be done several times, as having the outer face of the frame empty and the inner full can result in the wax foundation bending and even bursting if spun too vigorously!

Sometimes a super frame of stores is not completely capped with wax. The uncapped cells may contain nectar with too high a percentage of water to be regarded as honey. This leaves you with a tricky dilemma, as you need the final product to have an average water content of 18 per cent or less (20 per cent legally; 23 per cent for heather honey). Capped cells will likely be 16–18 per cent water, so some uncapped cells are acceptable, as all the honey will be mixed in the extractor. It is best not to extract these frames if the accompanying set of frames in this batch are also not very thoroughly capped with wax.

Of course, this problem will not occur if only frames that are well capped were selected for removal from the hive. It's reassuring to check the water content of some uncapped cells and the final honey using a hand-held refractometer. (There is a danger of fermentation in the jars if the water content is >19 per cent.)

Procedure

- Place the super to be extracted on a nearby surface that is clean – one that you don't mind honey dripping on to.
- Hold the frame with its longest dimension vertical and rest the other end in the uncapping tray.
- Using your uncapping knife, slice off the wax capping on the comb, trying to leave most of the honey beneath in the comb. The wax should be allowed to fall into your tray. Scrape a fork over any uncapped sections to ensure all

the comb is open. Reverse the frame and repeat the procedure on the other side. Once your tray is starting to fill with cappings, tip these into a spare clean bucket, along with any honey collected with it.

- Quickly place the frame in the extractor, to minimise the dripping of honey on the floor and table.
- Repeat until the extractor is filled. (If you have a helper, they can continue uncapping frames while you continue with the next stage.)
- Spin the extractor. You should be able to see and hear the honey spraying out to the internal circumference, and dribbling down to collect at the bottom. Take care not to go too fast, to avoid the comb being damaged.
- If using a radial extractor, continue until the frames appear empty – you will need to stop and check now and again that there is a minimal amount of honey left in the comb.
- If using a tangential extractor, only spin slowly at first, until half or more of the frame is empty on one side. Stop spinning and rotate the frame so that the full side is now facing out, repeat the spin until this side is empty, then rotate and spin again until the first side is empty. Repeat further if necessary.
- Once all frames in the extractor are empty, remove them and place back in the super box they came from.

The extractor will have enough volume at the base to hold at least a few frames' worth of honey, but eventually the spinning frames will drag in this honey. So you then need to open the valve on the extractor and let the honey pour through your strainers into a bucket. This will remove any wax and bee parts left in the honey. Sieving can be a slow process, and the sieves may become blocked – take care they don't overflow. You can multi-task and leave the valve open on the extractor so that it continuously drains while spinning – but you'll have to ensure the wobbling of the extractor does not result in it shifting across the kitchen when spinning. Once the filters are only draining slowly, they will need cleaning. Scoop out the wax that is blocking them and place with the wax cappings collected in the first steps.

Repeat the steps above, and you will quickly see your pool of filtered honey expanding – you can collect 10 to 12kg of honey from a full ten-frame super.

Once the bucket is nearly full, the lid should be added, then replaced with another clean bucket and the process continued. This full bucket should then be left for a day to allow entrained bubbles and wax to rise. In the meantime, perhaps immediately, a sample should be taken for a tasting ceremony. Fresh, buttered, wholemeal bread is a great medium of choice.

After the settling time, the honey can be transferred to jars if desired. Your bees' honey should be regarded as a premium product and deserves new jars and lids. Both need to be sterilised. To sterilise, wash the jars and lids in hot water, then

Filtering the honey from the extractor into a storage bucket

rinse and drain them upside down before placing on a tray in the oven, preheated to 140°C/Gas Mark 1 for 15 minutes.

Assuming the stored honey is in a plastic bucket with a tap, it's simple to place it overhanging the edge of a table, and crack open the tap a little to allow the stream of honey to flow into a sterile jar until full. Care is needed to do this slowly and avoid spillage. It's useful to place a bowl, or tissues, on the floor below to catch any missed dribbles. After a few jars you'll get the hang of closing the valve rapidly to avoid overfilling. The lid should be added quickly.

If you plan to sell your honey, you'll need to confirm the weight of it is at or above target. Most honey in UK is sold in the net metric weight equivalent of ½ lb, ¾ lb or 1 lb jars (that is 227g, 340g or 454g respectively).

Take care not to let the risen wax in the bucket pass into the jars. Inclining the bucket will maximise the clean honey recovered. If the bucket is nearing this point, other filtered buckets of honey can be tipped into it, and jar filling can continue the following day, once the honey has settled again.

Along with the honey, at the end of the process you will also have a bucket containing your collection of high-quality capping wax that is mixed with honey and other debris, and supers that contain sticky frames which are mostly empty of honey. These are dealt with overleaf.

There can be a substantial amount of honey mixed with the wax and debris. To collect this, it needs to be gravity filtered over 12 hours or more. To achieve this, it's simple to spoon the mixture into a muslin cloth suspended over a bowl and allow it to drip through overnight.

My preference is to store this honey in impressively large jars and use as 'home honey', for family consumption and for cooking. The wax remaining in the cloth, is very high quality and mostly made within the last few months. It can be rinsed in cold water, dried out (so it doesn't go mouldy), melted, and strained again, to form a block for use making candles etc. (see p.163).

Cleaning the supers

The best way to do this – and to remove any remaining honey – is to place the supers back on the hives they came from. If there is still a nectar flow occurring, they can just be placed above the queen excluder, and the bees will fill them further and place any precious drips of honey back into the comb.

The more likely scenario is that the flow will have finished and you want to dry up the supers and frames so they can be put into storage. In this case, the supers should be returned to the same hives, but placed above the crown board, with the Porter bee escapes removed. Over several days the bees will pass through these holes and remove any honey and place it into the frames below the board. Once complete, the bees can be removed from these supers (by putting the bee escapes back in place) and the now dry supers removed for storage.

Specific issues with extracting honey

If the source of most of the honey is from fields of oilseed rape, or heather, there are certain complications. For rape fields, the yield of honey can be very high, but the incoming nectar is high in glucose and will crystallise quickly, and even set in the comb if the supers are allowed to cool. To avoid this solidification as large crystals, it is best to extract immediately after the supers are removed and in warm conditions. Once placed in the jar the honey will crystallise and become very hard. So, prior to placing in jars, some beekeepers will carefully re-melt the honey and mix it with about 10 per cent of a soft-set honey at a cooler temperature, over several days, to create a similar soft-set product.

Heather honey has an unusual feature of being a 'thixotropic' liquid; i.e. it has a jelly-like consistency when still, but liquifies when stirred. This can make it difficult to extract from the comb. One way of achieving this is to use unwired comb and, when capped, cut from the frame prior to placing into a straining bag and compressing this comb using a honey press (very similar to an apple press).

If the beekeeper wishes to extract honey from comb produced in a top-bar or Warré hive, a honey press is also used.

Labelling

After admiring your beautiful line-up of sealed honey jars, you now need to consider labelling the jars. It's easy to design your own labels, or to order them online, but there are regulations that need to be followed if the jars are being sold in the UK. It's advisable to check the latest rules, but for honey from your own hives in 2024, here is a simplified guide:

- The label cannot be misleading and must include the word 'honey'. You cannot call it 'Lavender honey' unless you know for certain that 75 per cent of the honey originates from this plant. Many people use 'Local honey', or your region's name, such as 'Devon honey', as a description.
- The net weight of honey must be on the jars and in metric. You can also include the imperial equivalent, but it should be less prominent than the metric figure and not be in a larger font.
- For the jar sizes noted above, the minimum font size used for the weight must be 4mm. (If containers are larger than 1kg then the text should be larger than 6mm.)
- There must be a 'best before' date. Typically, this is 2 years from the date of extraction (though your honey will last much longer at its best).
- A 'lot number' has to be included for the batch of honey to enable the identification of the honey batch, e.g. L3. You can omit this if your 'best before' date includes a day, month and year.
- A traceable name and address must be on the label. It doesn't need to include every part of the address, but enough to allow you to be easily found.
- The country of origin of the honey must be stated – e.g. Produce of England.
- Any image used on the label must not mislead; e.g. you shouldn't show a picture of lavender if the honey isn't predominately from this flower.

There are other non-mandatory labels that can be added, such as a tamper-proof seal, a label advising about the natural process of granulation that may occur, and/or one stating that the product is 'Unsuitable for children under 12 months' (this is to protect from infant botulism).

A 'reference sample' jar of each lot number of honey should be retained by the beekeeper for several years, in case there is any problem with the honey that has been sold and further checks are needed.

Fresh wax cappings collected prior to extracting the honey

Beeswax

Humans have been using beeswax for medicine, food storage, beauty products and more as far back as the Stone Age. It was even used in dentistry as the first filling material to plug a cavity. Like honey, it's a precious resource and quite a remarkable product that bees make themselves when they are around a week old, once their special wax glands develop. The glands are located in a worker bee's abdomen and consumption of honey stimulates these glands to develop fully. The wax comes out in the form of small plates beneath the abdomen. Photos of this exist but we've never actually noticed it in a hive.

Each bee produces just 1g of wax in its lifetime, which is pretty extraordinary if you think of how much wax is in a hive – it's the work of a lot of bees!

How to harvest beeswax

When you harvest your honey, make the most of all the capped wax you've scraped from the super to unveil the honey. You can use this wax for all sorts of wonderful things, from making your own beeswax candles to lip balm.

Gather all the scraps of capped wax and place them on a wire cooking rack or other grid on top of the inner cover of the hive and then give the bees access to it through the feed hole. Surround the cappings with a spacer (called an eke) that goes between hive parts to create extra room for the bees to move around.

Leave the wax cappings in the hive for 2–3 days for cleaning. Then remove the wax which should now be cleaned of honey and give it a wash with cold soapy water. Drain the cleaned wax fully and pat it dry.

Alternatively, you can collect all the drained cappings in a bucket, then transfer it to a large pot (saved for this purpose) and cover it with clean water (rain water is recommended). Heat gently until melted: 68–70°C is optimal as it won't discolour the wax; then pour everything through an old kitchen sieve into another pot (again, not one you intend to use for cooking thereafter). Allow to cool slowly overnight.

The next day, remove the floating wax layer from above the syrupy liquid. Scrape away any bits from the bottom of the wax, or rinse off with boiling water. Your wax is then ready to use for candle making and more. You can either use it straight away or pour the melted wax into moulds or trays lined with greaseproof/baking paper. Leave it to set until fully firm and store the wax until you're ready to melt it down to make something fun (see recipes on pp.236–43).

You may find that some bee equipment suppliers will buy this wax from you, or swap it for product credits.

Propolis

Honeybees make propolis out of the resins they collect from deciduous trees such as cottonwood, birch, alder and poplar. As the trees bud, they exude these resins around the bud in order to protect it from fungi and diseases.

Foraging bees utilise their pollen baskets (corbicula) to carry globs of propolis resins back to the hive. Unlike with pollen, however, foragers require other bees within the colony to help them remove the sticky resins from their hind legs so it may be used by the colony.

The amount of propolis you can collect from one hive varies but you should be able to harvest enough to make a few tinctures and other medicinal products, as well as culinary treats. If you clean your hive well and use your hive tool or a metal scraper, you can increase your propolis harvest significantly.

Composition of propolis

The term propolis has Greek origins. In Greek, 'pro' means coming before or in front of, and 'polis' is the Greek word for a city or a body of citizens. Thus, propolis is what one could expect to find at the entrance to the city of the bees.

Beekeepers often observe that the bees will use propolis to restrict or narrow the entrance to the hive to make it easier to defend. They also use propolis as a building material, and as a way to sterilise and disinfect the cavity that contains the colony. This is because propolis is among the most powerful antimicrobial substances found in nature.

Over 240 compounds have been reported to have been extracted from honeybee propolis. While the composition of propolis will differ somewhat depending on which trees the bees gather the resins from, the typical composition tends to be 45–55 per cent resins; 25–35 per cent waxes and fatty acids; 10 per cent essential oils and other aromatic compounds; and around 5 per cent nutrient-rich pollen, another nutrient-rich substance. The essential oils include vanillin, which gives propolis its wonderful vanilla-like fragrance.

Extracting propolis

While you can buy traps that can be used to collect propolis, a kinder and more natural approach is recommended – to simply scrape excess propolis from the hive as and when you clean your frames. The propolis will be stuck in dark, burnt-amber-like clusters or nuggets around the edges of your frames.

Just use your hive bar tool or a knife to scrape off any propolis you don't think the bees need; i.e. if you're removing a frame, the bees will no longer need it. But don't take propolis from other parts of the hive, as the bees use it to fill gaps. It also provides antibacterial and antiviral functions to keep the bees and hive healthy.

To clean the propolis scrapings, place them in a roasting tin (not one you'll be using for cooking again) and cover with a 5–7cm depth of water. Place in a low oven heated to 100°C/Gas Mark ¼ for at least 2 hours, stirring often in order to release any wax that may be trapped within the mass of propolis. The melted wax, pieces of wood etc., will float to the surface of the water, while the propolis will stick to the bottom of the container.

After the tray is removed from the oven and cooled, the waxy layer on the surface of the water can be discarded and the water carefully poured off to reveal the reddish, resiny propolis mass beneath it. The container of propolis can then be frozen and, when the propolis is brittle, it can be chipped out of the container. The cleaned propolis pieces should be spread out on a sheet of paper or cardboard to dry before placing in a storage container.

There are various ways you can use propolis, including the preparation of medicinal tinctures and a rather fun propolis-infused oil which you can use to make the most delicious pine-meets-vanilla mayonnaise (see p.247).

Pollen: better left for the bees

While the carbohydrates in honey provide bees with energy, honeybees get all their vitamins, minerals, fats and protein from bee pollen. It's a vital food source for your colony and is what is fed to bee brood. Pollen also helps produce the royal jelly which is fed to the queen. Thus, harvesting pollen doesn't feel a comfortable option.

Most pollen is collected with the use of pollen traps, which are devices that fit over the entrance to a hive and contain openings just big enough for a returning forager to squeeze through. In the process of squeezing through the opening in the trap, the pollen carried on the hind legs of the bee is knocked off and falls through a screen into a drawer where it is collected by the beekeeper.

Bee colonies often modify their foraging behaviour and return with smaller pollen loads that can fit through the narrow opening of the pollen trap without being knocked off the hind legs of the returning bee, which does no favours to your bee colony.

Pollen is deliciously flowery and a full force of nutrients when it's super fresh, or stored as the bees house it, packed airtight in their wax cell 'jars'. The best way to obtain pollen yourself is by foraging pollen-rich flowers on a sunny day – just ensure you leave enough for the bees. Additionally, you can reap the healthy benefits of pollen by consuming raw honey, which has traces of pollen.

Rachel's Recipes

A teaspoonful of honey – a mere 5g – is the lifetime's work of a dozen bees. Considering their daily forage during their short life-span – each bee tapping nectar from hundreds of flowers across a vast (to them) and varied landscape and heaving it back to the hive to create those drops of gold – I, for one, have been taking honey for granted. And I don't seem to be alone. The average honey cake recipe calls for 250g of honey – the work of around 600 bees visiting more than a million flowers.

Reflecting on these calculations, I decided to take a different approach to the recipes that follow. To respect each drop of honey as the 'liquid gold' it is, I've opted for the 'less is more' philosophy. There's another reason for considered consumption. While honey may contain a host of nutritional benefits, including oligosaccharides (see p.142), enzymes, amino acids, vitamins and minerals, and have antiseptic and antibacterial properties (it's good for sore throats, skin conditions etc.), our bodies break down honey in a very similar way to sugar. While a little sweetness is no bad thing, indulging moderately is the wisest way forward.

The magic trick is to apply honey is a way that makes it matter more. Treating it as the 'icing on the cake' or a like a dazzling dusting of edible gold, makes it more relevant, not just in flavour, but in appearance. As tempting as it might be to drench dishes in honey, it seems disrespectful to bees; it also tends to steer us away from using the best local honey we can get our hands on.

Local honey may cost more – quite a lot more – but it's a case of quality over quantity, and supporting small-scale beekeepers is important to help preserve the global health of the honeybee. Many of the current threats to bees can be traced back to the commercialisation, you could even say exploitation, of bees – for both honey and pollination.

We also need to revisit how we cook with honey. To retain its flavour and goodness, it is best not to cook with it. When you heat honey above 37°C you lose valuable components, including antioxidants, which help to protect the body, and enzymes that give honey its antibacterial and antiviral properties benefits. If you heat honey to 100°C or above, you also destroy minerals and vitamins. Heating honey can also make it bitter in taste and more concentrated in sugars, changing its glucose structure, which can, again, have a negative impact on the body.

The good news is that, with a bit of rethinking and thoughtful application, you can still have your honey cake and eat it. The recipes that follow are rich in flavour and texture, and all have a golden kiss of luscious honey that sings – delighting your palate while offering health benefits.

While bees may produce more honey than they need, a generous act which allows us to indulge, we need to treat these creatures and the food they offer us with the utmost respect. So, venture into your kitchen with the best jars of honey you can get your hands on and embrace every spoonful.

Using honey in recipes

All of the recipes in this chapter illustrate skilful ways to use honey in your cooking without heating it detrimentally. Here are a few basic tips you can employ, however, when adapting recipes of your own.

- Adding honey to something warm, be it a cup of tea or coffee, is OK so long as the drink or dish is not piping hot; heating honey directly is best avoided.
- In baking, use fresh or dried fruit to sweeten the batter or dough of your cakes, muffins or breads, finishing them with a drizzle of honey once they are removed from the oven and have cooled slightly.
- You can fold honey through cooked dishes once they have cooled slightly, such as a batch of baked beans rich with tomato, or something flavoured with soy sauce or balsamic vinegar at the end, to balance the acidity and salt, and lend a richer finish.
- Honey is brilliant whipped into cream; blitzed with fresh fruit to make smoothies or a mousse; blended with soft cheese to make a honey-sweetened cheesecake; added to soup to lend floral notes; or used to sweeten nut butter or dairy butter to make a delicious spread.
- Freezing honey is also fun. It has a low water content, which means it won't freeze solid. However, as the temperature gets lower, your honey will become much more viscous and this can be used to culinary advantage. I've included a lovely lemon honey sorbet, as well as a honey ice cream recipe. In both cases, the texture is more akin to gelato – definitely a good thing. Unlike heating, freezing doesn't damage honey's goodness.
- Fermenting with honey is also a brilliant way forward. Not only do you retain the health benefits of honey, you increase the nutritional complexity, as well as evolving a whole new flavour spectrum.

Practical tips

- As a general guide, when using honey as a replacement for sugar in recipes, you will need less honey because it is almost twice as sweet as sugar. Replace 100g sugar with 50g honey.
- You should allow longer to incorporate honey into a mixture and beat more vigorously compared to sugar recipes.
- Consider the floral variety of the honey you're using, as it has the power to balance, enhance or impart some of its flavour to other foods.
- The easiest way to measure honey is to dip a measuring spoon into boiling water first, which helps the honey to slide more easily from the spoon rather than stick to it, helping ensure you don't lose a single drop.

Sweet chilli sauce

fermented with honey

Commercial chilli sauces are typically laden with sugar and often contain artificial colourings. This is a chilli sauce in its purest form; it also offers a boost of beneficial bacteria from the honey and the fermentation process. Even better, it's completely delicious and ridiculously easy to make. I love it with salt and pepper squid but it's equally gorgeous with halloumi, glazed over roast squash or drizzled on pizza.

Serves 4

75g fresh red chillies
1 garlic clove, peeled
75g honey
1 tsp freshly squeezed lime juice

You will also need

A sterilised lidded jam jar

Halve the chillies and remove the seeds if you prefer a milder sauce, otherwise leave them in. Slice or finely chop the chillies and garlic, then place in your jar. Add the honey and lime juice to the jar.

Put the lid on the jar, screw it on tightly and give the jar a good shake to mix the ingredients together.

Loosen the lid and leave at room temperature to ferment for at least a week or up to 1 month before using. It can then be stored in the fridge for up to 3 months.

Honey herb-infused vinegar

When I'm faced with a 'just about emptied' jar of honey, I make a honey-kissed herb-infused vinegar, choosing one of the herb options below. They're stunning and make a brilliant tonic – knock back a tablespoonful in a glass of warm water before bed and it could sooth indigestion and help you sleep better.

Honeyed herb vinegars are also a brilliant base for salad dressings. You can use them in cocktails, such as a Bloody Mary or Whisky Sour, too.

Makes 250ml

A few sprigs of your favourite herb (see below for medicinal benefits)
1 tbsp honey or an almost empty jar of honey
About 225ml apple cider vinegar

You will also need

A sterilised 250g lidded jam jar or bottle

Place the herbs in your jar or bottle with the honey and pour on the cider vinegar, adding more or less, as needed. Shake to help dissolve the honey and muddle the herb's aromatics with the vinegar.

The vinegar can be stored for up to a year in a dark cupboard at room temperature but can be used straight away.

Herb options and medicinal benefits

Bay Can help calm indigestion and treat migraines
Rosemary Immune-boosting and a cognitive stimulant, so good for mental focus and memory
Lemon verbena Rich in melatonin, which can help you sleep better
Mint Soothing for headaches and upset tummies
Thyme Great for sore throats and coughs
Sage May help to reduce bad cholesterol
Summer savory Considered an aphrodisiac by the Ancient Egyptians, Greeks and Romans, it is also great for digestion and sore throats
Basil Can protect skin from the effects of ageing
Oregano Contains antioxidants that can help protect cells from free radical damage, including cancer
Lovage Good for easing stomach pains
Tarragon Can help to improve sleep

Homemade honey vinegar

I've been making fruity versions of this for years, featuring apple cores and peelings, and more recently, banana or pineapple peels. All I do is pack the fruit scraps into a jar, pour on the solution below, cover the jar with a cloth and leave it to ferment for 2–3 weeks. The cloth allows the ferment to get air, which it needs, and it must be stirred often to prevent any growth on the fruit as it ferments. I then strain the fruit scraps out, pour the liquid into a fresh jar and leave it to ferment a further 4–6 weeks or until tangy like vinegar.

I've left the fruit scraps out of this recipe, focusing purely on the honey, but you can, of course, add them to your jar – they simply provide additional sugars as well as yeasts to help kick-start the ferment.

Makes 1 litre

1 litre filtered water
85g honey

You will also need

A sterilised 1.5 litre bottle or jar
Bottles for storage

Mix the water and honey and pour into your jar. Cover with a cloth and secure with a ribbon or rubber band. Leave to ferment for 2 months or until tangy and vinegar-like.

Bottle your honey vinegar and use as a luscious tonic – adding a splash of sparkling or tonic water and pairing it with seasonal fruit and herbs, or use in marinades or dressings, or in place of lemon juice for a finishing kick of acidity to balance a dish.

Hugh's honey and dukka
dip-a-thon

Simple, delicious, playful and endlessly adaptable – with different breads, honeys, oils and dukka blends – this idea came from Hugh when I asked him about his favourite honey recipes. It's the perfect start to any meal, or you can pair it with a seasonal salad to turn it into a simple lunch.

Serves 4–6

For the dukka

1 tbsp hazelnuts, roughly chopped or bashed
1 tbsp sunflower seeds
1 tsp coriander seeds
1 tsp fennel seeds and/or fennel blossom
1 tsp nigella (onion) seeds
A pinch of dried chilli flakes
A pinch of sea salt
A few twists of black pepper

To serve

A loaf of good bread
A dish of honey
A dish of good olive or local rapeseed oil

Set a small, heavy pan over a medium-high heat. Add the hazelnuts and toast for 2–3 minutes, shaking the pan regularly so they take on an even colour, and adding the sunflower seeds for the last minute or so.

Meanwhile, lightly crush all the spice seeds using a pestle and mortar – breaking rather than grinding the spices; leaving a few whole seeds is fine.

Add these to the pan of nuts and seeds, along with the chilli flakes, salt and pepper. Continue to heat for 2–3 minutes, moving or turning the mixture to prevent the seeds burning. Tip onto a plate and set aside to cool.

You can use the dukka immediately, or store it in an airtight container for a week or so. To serve, simply slice or tear your bread into pieces and set it out on a dish or board. Have the honey, oil and dukka in separate dishes. Get dipping! The preferred order is bread in oil, then honey, then dukka.

Feta-stuffed figs
with honey

You can enjoy this sumptuous and rather classic combination oven-roasted or blistered over an open fire then finished with honey. It is that perfect marriage of sweetness and salt that makes it so utterly moreish. Lovely with homemade flatbreads or pitta and salad.

Makes 8

8 ripe figs
200g feta
A few woody sprigs of rosemary, thyme or summer savory

To serve

A drizzle of good olive oil
A drizzle of honey

Preheat the oven to 200°C/Gas Mark 6 or heat up your barbecue.

Make a vertical slit in the middle of each fig and tuck in a 25g nugget of feta. Skewer, stuffed side uppermost, with a woody sprig of rosemary, thyme or savory.

Roast in the oven or blister over a smouldering barbecue until the figs are tender, caramelised and the feta is starting to melt.

Carefully transfer the figs to a platter or individual plates, trickle with olive oil and honey and serve warm.

Summer tomato salad
with a smoky honey dressing

After a day helping David Chalmers tend to the River Cottage bees, we ventured to the polytunnels down on the farm and were met with a colourful display of summery tomatoes – just begging for a simple yet stylish approach, and so this dish was born. It's now one of my favourite ways to eat summer tomatoes.

Serves 4

500g summer tomatoes
1 tbsp olive oil
1 tsp apple cider vinegar
1 tsp honey
A pinch of dried chipotle chilli flakes or hot smoked paprika
Sea salt and freshly ground black pepper

To serve

Fresh herbs and/or salad leaves (nasturtium leaves and oregano work well)

Slice larger tomatoes and cut smaller ones in half. Arrange on a large plate and dust with a good pinch of sea salt.

Whisk the olive oil, cider vinegar and honey together in a bowl and add the chilli flakes or smoked paprika and a good twist of black pepper. Trickle this dressing over the tomatoes.

Scatter summer leaves and/or herbs over the tomato salad and serve. It's delicious just with good bread and butter, but I also love it alongside freshly cooked lobster or crab.

Sardines with gooseberries
and a ginger honey teriyaki

A simple summery dish that works equally well with mackerel, trout or any other oily fish. I like to serve it with roasted fennel and a salad. The ginger honey teriyaki is the star here and it's amazingly versatile. Pair it with veg, fish and meat throughout the year – I love it with roasted squash in the autumn, or skewered mushrooms and celeriac chunks in winter.

Serves 2 as a main, or 4 as a starter

4 fresh sardines
A trickle of olive oil
200g gooseberries
Sea salt and freshly ground black pepper

For the ginger honey teriyaki

2 tbsp tamari or soy sauce
1 tbsp honey
1 tbsp apple cider vinegar
1 garlic clove, peeled and finely grated
1 tsp freshly grated ginger

To finish

Julienne of fresh ginger
A handful of fennel fronds

Preheat the oven to 220°C/Gas Mark 7. Line a baking tray with baking paper (this will prevent the fish from sticking to the tray).

For the ginger honey teriyaki, mix all the ingredients together in a small bowl.

Season the sardines all over and gloss with a little olive oil, then place on the prepared baking tray with the gooseberries.

Slide the tray onto the top shelf of the oven and cook for 2–3 minutes until the skin is lightly charred, then turn the fish over and cook for another 2–3 minutes until the other side is similarly charred but the fish are still nice and tender in the centre.

Brush the freshly cooked fish with the teriyaki sauce. Serve the sardines and gooseberries scattered with ginger julienne and fennel fronds.

Honey beans

Simple but satisfying, these homemade baked-style beans can be rustled up in an instant – even on the beach! Serve them up on buttered slabs of toast.

Serves 4

1 tbsp olive oil
2 garlic cloves, peeled and finely grated
2 x 400g tins of haricot beans (or 600g freshly cooked beans)
4 tbsp tomato purée
1 tbsp tamari or soy sauce
1 tsp apple cider vinegar
½ tsp smoked paprika
A pinch of ground mixed spice
1–2 tbsp honey, to taste
Sea salt and freshly ground black pepper

Set a frying pan over a medium heat. Add the olive oil, garlic and a pinch of salt. Swirl through and cook gently for a minute or until just golden.

Add the tinned beans with their liquid or the cooked beans with some of their cooking liquor. Simmer for 10 minutes or until warmed through, adding extra liquid if you've used tinned beans and there doesn't seem to be enough.

Swirl in the tomato purée, tamari or soy, cider vinegar, smoked paprika and mixed spice. Simmer for a further 5–10 minutes.

Take off the heat. Add enough honey to sweeten and season with salt and pepper to taste. Serve hot, piled onto buttered toast if you like.

Shallot tatin with orange
and a honey and nigella seed glaze

Naturally sweet shallots are slow-roasted in orange juice and butter until sticky and caramelised then capped with pastry and baked to crisp the pastry to golden perfection. Once the tart is cooked and turned out, a crowning glaze of honey is applied, then a dusting of oniony nigella seeds and sea salt for contrast.

Serves 4–6

500g shallots or small red onions (unpeeled)
250ml freshly squeezed orange juice
25g unsalted butter
A generous sprinkling of thyme leaves
Sea salt and freshly ground black pepper

For the rough puff pastry

200g plain flour, plus extra for dusting
A good pinch of sea salt
150g unsalted butter
2–3 tbsp ice-cold water

To finish

2 tbsp honey
A pinch of nigella (black onion) seeds
Extra thyme leaves

You will also need

A 20cm ovenproof frying pan

Preheat the oven to 200°C/Gas Mark 6. Put the unpeeled shallots or onions into a roasting tray and roast on the top shelf of the oven for 45 minutes, or until tender right the way through.

While the shallots are roasting, make the pastry. Mix the flour and salt together in a bowl. Cut 125g of the butter into 1–2cm cubes and gradually add to the flour, coating the pieces in flour as you do so. Rub the butter into the flour until it is almost at the breadcrumb-like stage (i.e. still with some bigger bits of butter). Add enough cold water to bring it together into a soft, silky (but not sticky) dough.

On a floured surface, pat the dough out and roll into a long rectangle, 1–2cm thick. Arrange a few very thin slices of the remaining 50g butter over the centre third of your pastry. Fold the bottom third up over the slivers of butter, then the top third down, as if folding a letter. Rotate the rectangle 90°. Roll out again and repeat this process five times, ending with a letter-folded piece of dough. Wrap the dough in a clean tea towel and pop in the fridge to firm up.

Tip the roasted shallots into a colander and rinse under cold water until cool enough to handle, then remove the skins from the root end (they should slip off easily). Trim off any papery tails, and if the shallots are on the larger side, halve them (root to tail).

Pour the orange juice into a 20cm ovenproof frying pan and heat until it reduces and thickens to a sticky caramel. Add the butter, thyme and a generous grinding of pepper.

Remove the pan from the heat and arrange all the shallots in the pan, nestling them snugly together in a single layer.

On a lightly floured surface, roll out the pastry to a large round and trim to a circle, 2cm larger all round than the pan. Drape the pastry over the shallots and tuck the edges down around the inside of the pan, so it's snugly blanketing the shallots. Bake in the oven for 25–30 minutes until the pastry is puffed up and golden.

Remove from the oven and leave to rest for 1 minute, then invert a serving plate over the tart. Holding the plate and pan tightly together, flip them over to turn the tatin out onto the plate. Finish with a trickle of your favourite honey, a good pinch of sea salt, a dusting of nigella seeds and extra thyme leaves.

Aubergine jackets
with spiced honey miso glaze

Miso and honey are delicious bedfellows, especially if you throw a kiss of spicy heat into the mix. While you can apply this umami-rich mix to all manner of dishes, it's especially good with slow-roasted or barbecued aubergines. That said, it's also heavenly with sea bass and roast tomatoes, or broccoli and buckwheat noodles.

Serves 4 as a side, or 2 as a main

2 large or 4 smaller aubergines
A trickle of olive oil
Sea salt and freshly ground black pepper

For the spiced honey miso glaze

3 tbsp miso
1 tbsp boiled water
1 tbsp freshly grated ginger
1 garlic clove, peeled and finely chopped
½ red chilli, finely chopped, or a pinch of dried chilli flakes (or to taste)
1 tbsp honey
¼ tsp Chinese five-spice powder

To finish

A handful of seasonal herbs (chervil is particularly good)
A pinch of freshly ground Szechuan peppercorns (optional)

Preheat the oven to 200°C/Gas Mark 6 or heat up your barbecue.

Halve the aubergines lengthways and score the cut surfaces on the diagonal at 1cm intervals, about 1cm deep; be careful not to pierce the skin. Now score across to give a criss-cross pattern. Sprinkle the scored surfaces with salt and pepper. Trickle with olive oil and rub to ensure the aubergine surfaces are evenly coated.

Place the aubergine halves, cut side up, on a baking tray and bake on the top shelf of the oven for 30 minutes or until tender and golden (on the verge of collapsing). Alternatively, cook on the barbecue, on both sides, until nicely coloured and tender.

Meanwhile, for the spiced honey miso, mix the ingredients together in a bowl to make a sweet, spicy, salty paste.

Once the aubergines are cooked, brush a thin layer of the miso paste over them. Serve scattered with herbs and sprinkled with a pinch of Szechuan pepper if you like.

Honey-glazed pork belly
with rhubarb and elderflower

Slowly cooked self-basting pork belly, rubbed with elderflower salt and roasted on a bed of rhubarb until everything is meltingly unctuous, this is one of my all-time favourite dishes. A shimmering honey glaze, ideally a summer, floral honey with a hint of elderflower.

Serves 6

1.2kg pork belly
4 rhubarb stalks
2 heads of fresh elderflower, plus extra to finish
2 tsp sea salt
A glass of sparkling elderflower or white wine
2 tbsp honey, plus extra to finish

Preheat the oven to 230°C/Gas Mark 8. Pat the pork belly dry and score the fat layer at 1cm intervals, then score across the cuts, to give a criss-cross pattern.

Slice the rhubarb on an angle, into 5cm pieces. Arrange in a single layer in the centre of a roasting tin. Sit the pork belly, fat side up, on top of the rhubarb.

Pick the elderflowers from their stems and mix with the salt. Rub the salt into the pork all over, and into the grooves of the scored fat.

Place the roasting tin in the oven and lower the setting to 170°C/Gas Mark 3. Roast the pork for 1½ hours. If the fat is not beautifully crisp once the cooking time is up, turn the oven to its highest setting or turn on the oven grill and cook for a further 10 minutes or until the fat has crisped up.

Lift the pork out of the tin onto a warm platter. Spoon the rhubarb into a separate dish. Let the meat rest in a warm spot for at least 20 minutes before carving. Meanwhile, make the gravy. Strain the pan juices and fat into a saucepan, place over a medium heat and add the sparkling elderflower or wine. Let it bubble up and reduce down, then remove from the heat. Strain to remove any bits, and whisk in the honey.

Before serving, warm up the rhubarb. Carve the pork and serve with the rhubarb and gravy, finishing with a trickle of honey and a sprinkling of fresh elderflowers.

Honeyed summer pudding

Summer pudding's juxtaposition of decadence and rustic simplicity makes it an endearing pud. It is far more than the sum of its parts (stale bread and sweetened fruit), and a brilliant way to use a glut of homegrown soft fruit. Adding honey takes it to another level. And the idea that your bees might have foraged on the blossom from your fruit before it swelled to its perfect ripeness is pretty magical.

Serves 6

1.2kg mixed summer soft fruit (such as blackberries, raspberries, redcurrants and strawberries)
4 tbsp honey
7 slices of day-old white bread, ideally from a square, medium-cut loaf

You will also need

A 1.2 litre pudding basin or 6 large ramekins (200ml capacity)

Wash the fruit and pat dry on kitchen paper; if using strawberries, keep them separate. Put the honey and 1 tbsp water into a large bowl and whisk to combine and form a thick syrup (use warm water or set over a pan of hot water if using set honey). Fold the fruit through the mix, fairly vigorously to combine and release some of their juices. Tip into a sieve set over a bowl to catch the juice.

Cut the crusts off the bread. For a large pudding, cut 4 slices in half, slightly on an angle, to give two wonky rectangles each; cut another 2 slices into 4 triangles each; leave the final piece whole. For individual puds, cut 6 bread discs to line the base of the ramekins, rectangular pieces to line the sides and 6 discs to cover the tops.

To build the pud(s), dip the whole piece of bread (or 6 discs) into the juice for a few seconds just to coat, then use to line the bottom of the basin (or ramekins). Now dip rectangular pieces one at a time into the juice and press around the sides of the basin (or ramekins) so they fit together neatly, alternately placing wide and narrow ends up and trimming as necessary. Spoon the softened fruit into the bread-lined basin (or ramekins).

Dip the bread triangles (or discs for individual puds) in juice and place on top of the pudding(s). Trim away any overhanging bread with scissors. Save any leftover juice for serving. Put a side plate on top (use saucers for ramekins) and weight down with heavy weights or something similar. Chill overnight, or for at least 6 hours.

Invert a serving plate over the pudding(s), hold the bowl and plate tightly together and flip over to turn out the pudding. Serve with the saved fruit juice and cream.

Blackberries and honey cream

This delightful, easy pud is very versatile – you can swap out the blackberries for other seasonal fruit. It's perfect with any ripe summer berries or stone fruit, but also delicious made with roasted apples, pears or quince in the autumn. In the spring, try using sliced rhubarb poached in orange juice with a couple of star anise and trickled with honey once cooled.

Serves 4

500ml whipping or double cream
4 tbsp honey
350g blackberries (or other seasonal fruit)
Lavender (or other edible flowers or seasonal herbs), to finish (optional)

In a large bowl, whip the cream until thickened enough to hold its shape. Ripple half of the honey through the whipped cream.

Trickle a little honey into each of 4 glasses then layer the whipped cream with the fruit in the glasses. Trickle with the remaining honey and finish with a few lavender flowers if you have them.

Chill until ready to serve or eat straight away.

Tip Vary the finishing touch, as well as the fruit, with the seasons. Rose petals are stunning with a strawberry fool, while a pinch of saffron threads whipped into the cream rather than sprinkled on top is lovely with a quince fool.

Chamomile honey jelly

A stunning, soothing jelly that celebrates the excellent pairing of honey and chamomile. It uses the setting agent agar agar, which is made from seaweed, so it's a vegetarian-friendly pud.

Serves 4

500ml freshly brewed chamomile tea (made by infusing 2 chamomile tea bags or 2 tbsp loose leaf chamomile in 500ml hot water from the kettle)
2 tbsp agar agar flakes
3–4 tbsp runny honey, to taste, plus extra to serve
Fresh or dried chamomile flowers, to finish

You will also need

4 jelly moulds or tea cups

Strain the tea through a fine strainer into a saucepan and sprinkle in the agar agar flakes. Warm the mixture over a medium heat until the agar agar dissolves, whisking vigorously a few times to encourage it to do so.

Pass through a sieve into a bowl to strain out any lumpy bits of agar, pressing the residue in the sieve with the back of a wooden spoon to extract as much as possible. Whisk again, then set aside to cool to room temperature.

Taste the liquid jelly and stir through enough honey to suit your taste-buds. Pour into jelly moulds or tea cups and pop into the fridge for about 2 hours to set.

Top the individual jellies with a dusting of fresh or dried chamomile flowers and a final trickle of honey to serve.

Honeyed hop panna cotta

The cone-shaped flowers of the hop plant are in season in the autumn. In their dried form, their most familiar use is as a flavouring and bittering agent in the brewing of beer, but they also lend floral, fruity notes to herbal teas and can be used fresh to flavour dishes such as panna cotta. I like to serve these panna cottas with seasonal fruit such as poached damsons or quince, on the side.

Serves 4

500ml single cream
50g hops (ideally fresh, but dried can be used)
2 sheets of leaf gelatine (or 2 tbsp agar agar for a vegetarian version)
100g natural yoghurt
3 tbsp honey, plus extra to serve

Pour the cream into a saucepan and set over a medium-low heat. Stir the hops through the cream and warm through for 5 minutes, but don't let it bubble. Take off the heat, give it another stir then set aside to infuse for 30 minutes.

Soak the gelatine, if using, in a shallow dish of cold water for 2–3 minutes until soft and pliable.

Reheat the infused cream almost to a simmer, the strain out the hops. Squeeze out the excess water from the gelatine then immediately add to the warm cream, stirring as you do so. Continue to stir until the gelatine is fully dissolved. (Or if using agar agar, add it straight to the warmed cream, stir until dissolved then strain through a sieve, pressing it through with the back of a wooden spoon.)

Leave the cream mixture to cool to room temperature, stirring from time to time. Stir in the yoghurt and honey until well combined then pour the mixture into serving glasses. Chill in the fridge for at least 4 hours, until set.

To serve, finish with a trickle of honey and serve with seasonal fruit on the side.

Variation

Honeyed rose panna cotta Swap the hops for fresh (or dried) rose petals. Delicious with strawberries.

Honeyed chocolate pots

You can either serve these in little pudding pots or let the chocolate set and then roll it into truffles, coating them in crushed nuts or a dusting of cocoa (see below).

Makes 4 pots or 12 truffles

100g dark, bitter chocolate (at least 85 per cent cocoa solids)
100g double cream or coconut milk (for dairy-free truffles)
1 tbsp melted butter or coconut oil
2–3 tbsp honey, to taste
A pinch of sea salt

Finely chop the chocolate and place it in a heatproof bowl.

Gently warm the cream or coconut milk in a small pan over a low heat until just steaming. Pour onto the chopped chocolate and leave to melt – don't stir until all or most of the chocolate has melted. If the chocolate isn't fully melted, set the bowl over a pan of gently simmering water, making sure the base isn't in contact with the water. It is important to let the melting happen slowly and gently, as heating chocolate too quickly can cause the cocoa butter to separate from the cocoa solids.

Once the chocolate is all melted, gently fold in the melted butter or coconut oil and enough honey to sweeten the mixture to your liking.

Pour into individual pots and refrigerate until ready. The chocolate pots will keep in the fridge for up to 3 days; you can take them out 10 minutes or so before eating to soften slightly.

Variation

Chocolate truffles If you want to make these, pop the bowl of melted chocolate mixture in the fridge to set. To shape the truffles, scoop out teaspoonfuls of the set mixture and roll into smooth balls. Now roll the balls in crushed nuts or cocoa powder to coat all over. Refrigerate until ready to eat.

Banana mousse
with honey and nut butter

The perfect energy boosting treat for little ones and adults alike! You can serve it up as a pudding or snack, or slather it on toast, or pipe it into a doughnut. It's endlessly versatile and super-delicious.

Serves 2

2 ripe bananas
2 tbsp nut butter (I like cashew or peanut), or tahini
1 tsp honey, or to taste, plus extra (optional), to serve
A pinch of sea salt (optional)

Peel the bananas. Using a handheld stick blender or jug blender, simply blend all the ingredients together until smooth.

Divide the mousse between serving glasses and serve at once, with a little more honey swirled through if you like.

Honey roast pears

When pears are slow-roasted, their natural sugars caramelise and offer the most heavenly sweetness. All you need is a final trickle of honey to take them to the next level, and maybe a nice dollop of clotted cream… even, perhaps, a warm chocolate brownie on the side.

These pears are also scrumptious with crêpes, or porridge enriched with cream and a pinch of sea salt.

Serves 4

4 large or 8 smaller pears
2 tbsp butter
A pinch of ground cinnamon and/or freshly ground cardamom
100ml brandy, wine, water or spiced tea
1–2 tbsp honey, to serve

Preheat the oven to 180°C/Gas Mark 4. Put a baking dish or ovenproof pan into the oven to heat up. Meanwhile, peel and halve the pears, then scoop or cut out the core and seeds.

Take the warmed dish or pan out of the oven. Add the butter and swirl it around until it is melted.

Add the pears to the pan, sprinkle with the spice and turn to coat in the melted butter. Add the brandy, wine, water or spiced tea. Place in the oven and roast for 30 minutes or until the pears are sticky, golden and on the point of collapsing.

Trickle the pears with the honey and serve warm or cold.

Honey custard

Custard is typically made by whipping egg yolks with sugar, then stirring in gently warmed cream and simmering until it is thick enough to coat the back of a spoon. Instead of sugar, this version is sweetened with honey, which is folded in at the end, to avoid overheating it.

The custard is delicious with seasonal fruit and puddings, or you can freeze it to make a phenomenal ice cream (see below). You can have fun using different honeys, too. Ice cream flavoured with white-truffle-infused honey is spectacular – especially served with slow-roasted figs and a touch of aged balsamic. It's also rather special made with chestnut honey and partnered with a warm chocolate brownie or a freshly baked apple cake.

Serves 4

4 egg yolks
400ml single cream
40–50g honey, to taste

Put the egg yolks and cream into a heatproof bowl and whisk to combine. Set the bowl over a pan of simmering water, making sure the water doesn't touch the base of the bowl, and cook over a low heat for about 10 minutes, stirring constantly, until the custard thickens.

Strain the custard through a sieve into a bowl and whisk in the honey to taste. Serve with seasonal fruit.

Variation

Honey ice cream Make the custard using 50g honey (it will taste less sweet once frozen). Allow to cool completely then pour into a shallow freezerproof container (a metal one will help it to freeze faster). Cover and freeze for at least 4 hours or until solid – no need to whisk or churn as the honey helps prevent ice crystals from forming. Scoop the ice cream into serving bowls and finish with a trickle of honey. Serve straight away, as it melts rather quickly.

Roasted squash custard tart
with honey

A beautiful evening of pumpkin carving and feasting around a bonfire at River Cottage HQ inspired this recipe. It is a classic Halloween or harvest festival treat, especially in America where I grew up, and it's delicious served with a hot drink. I've given the tart a healthy makeover by leaving out the sugar, sweetening the custard base with dates and apple juice instead and glazing the finished tart with honey. It works a treat and adds the perfect layering of sweetness with honey's classic acidic undertone. I love to use chestnut honey here.

Serves 6

For the shortcrust pastry
200g plain flour
A pinch of sea salt
100g unsalted butter, cut into small cubes
1 egg, beaten
A little ice-cold water

For the filling
500g roasted squash (butternut, Crown Prince, kabocha or onion squash), see recipe method
2 eggs
100g soft pitted dates
2 tsp ground cinnamon
½ tsp ground ginger
⅛ tsp ground cloves
A good grating of nutmeg
A pinch of sea salt
250g single cream
100ml apple juice

To finish
2–3 tbsp honey

You will also need
A 23cm tart tin

Preheat the oven to 200°C/Gas Mark 6. For the filling, if you haven't roasted your squash already, the best way is to halve it (around the girth), scoop out the seeds and roast the halves, cut side up, for 1 hour or until the squash is meltingly tender (no need to season or oil). Leave to cool slightly for 10 minutes or so, then scoop out 500g flesh.

Place the squash in a blender or food processor and add the eggs, dates, spices, salt, cream and apple juice. Blitz until smooth and creamy, then transfer to a bowl, cover and chill in the fridge while you make your pastry.

To make the pastry, mix the flour and salt together in a large bowl and toss through the butter cubes. Using your fingertips, rub the butter into the flour until the mixture resembles breadcrumbs. Add the beaten egg and mix with a table knife or fork, then sprinkle in enough cold water, 1 tbsp at a time, to bring the pastry together into a ball.

Wrap the pastry in baking paper and place in the freezer to firm up for 15 minutes.

Remove the pastry from the freezer. Grate it into the 23cm tart tin set on a baking sheet. Press the grated pastry into the tin as evenly as possible, fully covering the base and the sides, right up to the rim. Line the case with a sheet of baking paper and add a layer of dried beans or baking beans. Bake in the oven for 10 minutes, then remove the paper and beans and return to the oven for 5 minutes or until the pastry base is golden. Remove from the oven and leave to cool for 10 minutes.

Spoon the filling into the pastry case and smooth the surface. Bake for 10 minutes, then lower the oven setting to 180°C/Gas Mark 4 and bake for a further 40–45 minutes, or until a knife inserted in the centre comes out clean.

Transfer the tart to a wire rack to cool and drizzle the honey over the surface while it is still warm. Leave to cool completely, about 2 hours, or refrigerate and eat within 2 or 3 days.

Apple baklava

Much of the baklava you can buy these days is made with glucose syrup or sugar in place of the honey. In this recipe, the honey plays the starring role but I've also lightened the baklava by adding grated apple to the nut mixture, which lends sweetness and moisture, but also a luscious autumnal flavour. Filo pastry is suggested for convenience, but this also works well with homemade rough puff pastry (see p.186) as pictured overleaf, if you roll it out thinly.

Makes about 15 pieces

For the filling
2 apples
300g shelled nuts (cob nuts, hazelnuts and/or walnuts), lightly toasted
1 tsp ground cinnamon
A pinch of sea salt

For the pastry
220g ready-made filo pastry
100g butter, melted

For the syrup
150g honey
1 tbsp apple juice (squeezed from the apples for the filling, see recipe)

You will also need
A 25cm square baking tin (or one with similar dimensions)

Preheat the oven to 180°C/Gas Mark 4

Peel, core and coarsely grate the apples then tip them into a sieve over a bowl and squeeze firmly to extract 1 tbsp juice; if you have more than this quantity, simmer to reduce down; set aside for the syrup.

For the filling, put the grated apples into a bowl (they should feel quite dry). Grind half of the nuts quite finely (but not as fine as ground almonds). Chop the rest of the nuts and add to the grated apples with the ground nuts, cinnamon and salt. Mix thoroughly to combine and set aside.

Trim the filo pastry to fit the dimensions of your baking tin – you will probably need two sheets to create each layer. Brush the base of the tray with melted butter, then add a layer of filo and brush with more butter.

Sprinkle a very thin layer of the apple and nut filling over the filo in the tin as evenly as possible; do not press it down too much. Cover with another layer of filo and brush with butter. Repeat until you have four or five layers, finishing with a layer of filo.

Score the baklava into squares by cutting just through to the filling in parallel lines. Bake for around 40 minutes until the top is crisp and a rich golden brown colour. Remove from the oven and leave to cool.

Meanwhile, for the syrup, put the honey and apple juice into a jug and whisk to combine. Pour the syrup evenly over the pastry. Leave to cool completely, for at least an hour and preferably for a few hours.

Remove the baklava from the tin and cut into squares. Use a metal dough scraper or a long knife, so you can just press down instead of pulling through.

Serve the baklava on its own or with a dollop of yoghurt, with coffee or tea – it's particularly good with fresh mint and lemon verbena tea. The baklava will keep covered at room temperature for up to 2 days.

Variation

Quince baklava Swap the apples for quince: you want approximately 200g grated fruit. To draw out the flavours, swap the apple juice for quince juice and have a play with the spicing, by adding ½ tsp ground cardamom to the fruit and 2 tsp orange blossom water to the honey and juice. For a Christmassy touch, use freshly roasted chestnuts as your nutty base.

Lemon honey sorbet

A scrumptious, light, sweet treat to end a meal or refresh your palate that is also magically soothing for sore throats.

Makes 500g

Finely grated zest of 1 lemon, plus 350ml freshly squeezed lemon juice (about 6 lemons)
150ml honey
A pinch of ground cinnamon and/or thyme leaves and flowers to finish (optional)

Whisk the lemon zest, lemon juice and honey together in a bowl. Transfer to a shallow freezerproof container, cover and freeze for at least 4 hours or overnight.

To serve, scoop the sorbet into bowls and sprinkle with a little cinnamon and/or thyme, if you like. Eat straight away.

This sorbet keeps well in the freezer for up to 6 months.

Honey halva

This tastes like halva but takes just minutes to make. It's rich in B vitamins and sweetened only with honey, making it a nutritious after-dinner treat, or the perfect biscuit alternative for mid-morning coffee.

Makes about 32 bite-sized pieces

200g light tahini
3 tbsp honey
1 tbsp ground cinnamon
½ tsp freshly grated ginger (optional)
Finely grated zest of 1 lemon
Sea salt flakes, to serve

You will also need

A small shallow baking tin (about 15cm square)

Line your tin with baking paper, leaving plenty of excess overhanging the sides (to fold over the top of the halva).

Mix all the ingredients together in a bowl until evenly combined. Then, using your hands, knead the mixture until it comes together to form a fudge-like paste. Press the mixture into the prepared tin and smooth the surface. Fold the overhanging baking paper over the top to cover.

Place the halva in the freezer for about 30 minutes to firm up. It is then ready to eat, or it can be kept in the fridge for up to 2 weeks.

Cut the halva into small squares and sprinkle with sea salt flakes to serve.

Honeyed carrot cake

Reinvented to replace the sugar with dried fruit and feature honey as a sweet, sticky, golden glaze, this is now my favourite carrot cake recipe. Coriander honey is a lovely match for this, or opt for a honey made on a small scale by a beekeeper near you. Squash works nicely as an autumnal swap for the carrots and raisins can replace the dates.

Serves 8

2 large eggs
200g natural or Greek-style yoghurt
100ml olive oil, plus extra to grease the tin
150g pitted dates
Finely grated zest of 1 orange and 1 tbsp juice
2 tsp baking powder
1 tbsp ground mixed spice
1 tbsp freshly grated ginger
150g carrots, coarsely grated
100g spelt flour or wholegrain flour
75g ground almonds
50g walnut halves (optional)
3 tbsp honey

You will also need

A 900g (2lb) loaf tin

Preheat the oven to 180°C/Gas Mark 4. Lightly oil the loaf tin and line the base with baking paper.

Put the eggs, yoghurt, olive oil, dates, orange zest and juice, baking powder, mixed spice and grated ginger into a food processor or blender and blitz to a smooth purée. Transfer to a large bowl.

Carefully fold the grated carrot into the mixture, then gently fold in the flour and ground almonds. Transfer the mixture to your prepared tin and gently level the surface with the back of a large spoon. Top with walnut halves, if using.

Bake in the centre of the oven for about 1 hour until golden and set in the middle. To test, insert a skewer or small knife into the middle of the cake; if it comes out clean, the cake is cooked.

Place the cake tin on a wire rack and leave the cake to cool until just warm – or completely – before removing from the tin. Drizzle the honey evenly over the surface to glaze.

Tip If your dates are dry (i.e. not pitted Medjool), they will benefit from being simmered in orange juice to soften. Use the rest of the juice from the orange in the recipe and simmer the dates for 10 minutes or so.

Honey buns

My favourite milk bread, rolled into cinnamon-packed buns and baked until golden, then glossed with honey and butter as you take it from the oven. I can't think of many things I'd rather eat on a Sunday morning with a pot of coffee.

Makes 12–15

For the dough
250ml tepid milk
250g activated sourdough starter, or 7g fast-action yeast
500g plain flour, plus extra for dusting
A big pinch of salt
100g butter, melted
1 tsp ground cinnamon
2 large eggs, beaten

For the filling
50g butter, melted
1 tbsp ground cinnamon

For the glaze
100g honey
100g butter, melted

You will also need
A 26cm round baking tin

For the dough, mix all the ingredients together in a large bowl until smoothly combined. Knead well for about 3 minutes to make a stretchy dough. Place in a clean bowl, cover with a damp cloth and leave to rise at room temperature until doubled in size: 8–12 hours or overnight if using sourdough; 2 hours if using yeast.

Generously dust a clean surface with flour. Turn out the dough onto the surface and roll/stretch out to a rectangle, roughly 25 x 40cm.

For the filling, mix the melted butter with the ground cinnamon. Trickle or brush over the rectangle of dough. Roll the dough up, starting from one long side, to make a nice, tight cylinder.

Slice the cylinder into 12–15 equal pieces. Turn each piece of dough on its side so a spiral face is upwards and nestle the pieces into the baking tin. Cover with a tea towel and leave somewhere warm for about 1 hour, or until the buns are risen and puffy. Meanwhile, heat the oven to 180°C/Gas Mark 4.

Bake the buns for 20–25 minutes until they are risen, golden and firm to the touch. Meanwhile, mix the honey and melted butter together for the glaze.

Trickle the honey glaze over the buns as you take them out of the oven. They are best eaten warm from the oven, but will keep in an airtight container for 2–3 days at room temperature – reheat before eating and they'll still be delicious.

Honey tahini flapjacks

A healthy, raw, quick and easy take on the traditional flapjack that's rich in tahini and is sweetened only with honey and fruit. Dazzled with a magical dusting of Iranian flavours, it's a great energy-boosting treat.

Makes 12–15

250g oats
50g chopped nuts or a nut/seed mix
200g light tahini
1 ripe banana, peeled
100g soft pitted dates (or other dried fruit, such as figs or apricots)
1 tbsp ground cinnamon
8 cardamom pods, seeds extracted and ground
50g honey
50g butter or coconut oil, melted
2 tsp orange blossom water (optional)
A pinch of sea salt

You will also need

A 24cm square baking tin or one with similar dimensions (ideally that has a removable base)

Preheat the oven to 180°C/Gas Mark 4 and put a large baking tray on the middle shelf to heat up.

Scatter the oats and nuts on the hot baking sheet and bake for 10–15 minutes or until lightly toasted. Remove and set aside to cool. (You can skip this step and leave the oats and nuts raw, but the flavour is improved with a light toasting.)

Using a handheld stick blender or a jug blender, blitz the tahini, banana, dates, ground spices, honey, butter or coconut oil, orange blossom water, if using, and sea salt together until smooth. Transfer to a bowl (if you need to) and fold through the toasted (or untoasted) oats and nut (or nut/seed mix).

Line the base and sides of your baking tin with baking paper. Pack the flapjack mixture into the tin, pressing it down firmly and smoothing the surface.

Chill in the fridge for 2 hours or until firm. Remove from the tin and cut into squares. Enjoy straight away or keep the flapjacks in a sealed container in the fridge and eat within a week.

Beeswax-fermented fruit

Former River Cottage chef Mark McCabe inspired this idea: a keen fermenter, he was experimenting with fruit ferments at The Ethicurean restaurant near Bristol. I decided to have a play and found it deliciously fun, experimenting with a variety of fruits. The wax protects the fruit, giving it a safe, sterile environment to ferment.

Strawberries are my favourite, and they're one of the fastest and easiest fruits to ferment, too. Their flavour is intensified by the fermentation process so they become perfect summer flavour bombs to add to desserts – they're particularly stunning with my honey ice cream on p.205 and honeyed rose panna cotta on p.197. They are also lovely added to cocktails – or just eaten on their own as a fun finale to a meal.

Makes 12

12 strawberries with long stems
2 tbsp natural yoghurt
200ml water

You will also need
60g beeswax

Wash the strawberries and carefully remove any leaves from the stems but keep the stems in place.

Mix the yoghurt with the water. Dip the strawberries in this mixture (this helps kick-start the fermentation process). Place on a wire rack, pop in the fridge and leave for 30 minutes – 1 hour to dry completely.

Grate the beeswax into a heatproof bowl and set over a saucepan of simmering water, making sure the water doesn't touch the base of the bowl. Let the beeswax melt fully, stirring often.

Holding it by the stem, dip each strawberry in the melted wax to coat it completely, right up to the stem. Lift the wax-coated berry out and hold it above the bowl to let any excess wax drip off – it will start to cool and set as you do this.

Place the coated berry on the wire rack to finish cooling and setting. Repeat with the remaining berries and wax. Once all the berries are coated and on the wire rack, ensure they're still fully covered in wax; i.e. make sure there are no cracks or patches where the wax has dripped off, re-coating if necessary.

Leave the berries to ferment in a cool place out of direct sunlight for a week.

You can either cut into them or simply peel off the wax – do this over a bowl to catch the juices.

Christmas quincemeat
fermented with honey

While making a carrot-cake-inspired honey fermented jam of grated carrots mixed with sweet spices, dried fruit and citrus, someone remarked that you could apply the idea of fermenting ingredients in honey to Christmas mincemeat, leading me to create this recipe.

You can spoon this mixture into pre-baked pastry cases or simply dollop it onto drop scones with cultured (or brandy) cream, or enjoy it with yoghurt or alongside a panna cotta (see p.197), or just have it on toast.

Makes about 200g

½ quince or 1 small apple
1 tbsp cob nuts or blanched almonds
1 tsp freshly grated ginger
½ tsp ground mixed spice
½ tsp ground cinnamon
A dusting of freshly grated nutmeg
1 tbsp pitted chopped dates
1 tbsp currants
1 tsp grated orange zest
1 tsp grated lemon zest
100g honey

You will also need

A sterilised 200g (or slightly larger) jar

Coarsely grate the quince or apple, avoiding the core, and place in a bowl. Thinly slice the cob nuts or almonds and add to the bowl with all the remaining ingredients. Mix together thoroughly.

Spoon the quincemeat into your jar, ensuring you leave a little headspace, and secure with a lid.

Leave to ferment at room temperature for 2 weeks, shaking or stirring it every day or so, to ensure you don't get yeasts (which can lead to mould growth) settling on the top.

Once fermented, you can either leave the quincemeat to ferment further, stirring it every so often (as above), or store it in the fridge. It will keep happily for weeks, if not months.

Elderberry tonic
fermented with honey

Elderberries ripen in the hedgerows as we edge into September and through October. Not only does their inky offering please the palate, but these vitamin-C rich berries are also brilliant for boosting our immune systems ahead of the sniffle season. This is a honey fermented take on a classic elderberry tonic, with fresh ginger and rosemary added for extra flavour and goodness.

Makes 150ml

75g elderberries
1 sprig of rosemary
3 slices of ginger, each 1cm thick
3 cloves
75g honey

You will also need

A sterilised 200g (or slightly larger) jar
A sterilised bottle for storage

Put the elderberries, rosemary, ginger and cloves into your jar – you want some headspace in the jar to give the honey mixture oxygen, which helps the fermentation process. Pour the honey over the elderberries and aromatics, then mix through.

Put the lid on and store at room temperature, away from direct sunlight for at least 2 weeks, or up to 3 months. Give the jar a shake every day or two to help keep the berries covered and mixed in with the honey, opening the jar for a while after you do so to enable the ferment to get more oxygen. You're aiming for the berries to swell and bleed into the honey. As they ferment, the berries effectively 'cook', making them palatable (raw elderberries can cause stomach upset).

Once the elderberries have plumped up and their juices have started to bleed into the honey, and the mixture looks thinner and bubbly, press and scrape everything through a fine-meshed sieve or squeeze through a large square of muslin into a bowl. Decant the syrup into a bottle.

You can use your elderberry tonic as a cough syrup or take it by the tablespoon to help strengthen your immune system. Or enjoy it with more frivolity by drizzling it over pancakes or pairing it with English fizz for a hedgerow twist on Kir Royale.

Mead

Mead or honey wine has been around since at least 7000 BC. The first written description is in the *Rigveda* (ancient collection of Vedic Sanskrit hymns); Aristotle references mead in *Meteorologica* and Pliny the Elder in *Naturalis Historia*; while the mead hall was at the heart of the epic Anglo-Saxon poem *Beowulf*.

Making this historic drink, however, doesn't need to be epic or involve vast buckets of honey, or specialist brewing buckets, for that matter. I've distilled both portions and process down into such an easy formula that you can probably lay the book down now and get started.

Makes 750ml

400g honey
350ml water
Juice of 1 lemon
1 tbsp raisins
A slice of ginger, 1cm thick

You will also need

Two sterilised 1 litre bottles

Whisk the honey, water and lemon juice together in a jug. Tuck the raisins and ginger into one of the sterilised bottles and pour in the honey mixture. Put the lid on the bottle and give it a good shake.

For the first stage of fermentation – when the yeasts in the raisins and ginger are consuming the natural sugars in the honey and producing gases – you need to create an airlock – one that helps keep oxygen in, but allows the gases that are building up in the bottle to escape.

You could just 'burp' or gently open the lid just enough to let the building pressurised gases escape from your bottle a few times a day – most of the bubble action happens in the first week. Or you can buy an airlock and rubber stopper with a hole drilled in for the airlock to fit snugly inside – there arc small airlocks made to fit thin-necked 1 litre bottles.

Alternatively, you could fit a small balloon around the neck of your bottle using a rubber band to secure it – this is fun as the gas from the fermentation process will inflate the balloon. Once you have a day without the balloon inflating, the first stage of fermentation is complete.

After 7–10 days the bubbling action should have stopped and you should have a layer of yeast at the bottom your bottle – these are dead yeasts that you don't want in your final brew.

The next step is to syphon off the honeyed liquid from the dead yeast. Traditional brewers will place a special pump in the bottle to draw up the liquid off the yeast and transfer it to your other sterilised bottle. However, if you don't want to purchase this extra kit you can gently and slowly pour the liquid through a funnel into a fresh sterilised bottle (or pour it into a jug then into a new bottle), being careful to leave the dead yeast behind.

Once you have your brew in a fresh bottle, cap it again with an airlock or a tight-fitting balloon and ferment for a further 2 months. At this stage, the liquid should be clearer and you can transfer it again to a clean bottle, discarding any more yeasts. It is now ready to drink, but will improve if aged in the bottle further.

Seven years is said to be the magic number for maturation. I've yet to harness my patience to wait that long, but if you can leave it for at least a year, your taste-buds will be awarded. A valid motto: if it tastes good, drink it!

Variation

Cyser A cross between cider and mead, this is a heavenly alternative to mead. Simply swap the water for apple juice, preferably freshly pressed, and brew as above.

Jun

Jun is a fermented tea made with a culture called a scoby (Symbiotic Colony of Bacteria and Yeast), which looks a bit like a jellyfish. It is effectively the same as kombucha but the scoby for jun is strong enough to withstand the antibacterial effects of honey, whereas a kombucha scoby often cannot.

This is a simple, straight jun recipe, which can vary dramatically depending on the type of honey you use. The fruity variations opposite are best made with milder, more subtle-flavoured honeys.

Makes 1 litre

4 tea bags or 4 tbsp loose leaf tea (black, green or rooibos tea works best)
1 litre boiling water
75g honey
1 jun kombucha scoby

You will also need

A 1.5 litre jar
A sterilised bottle (or bottles) for storage

To brew your tea, put the tea bags or loose tea into a jug and pour on the boiling water. If you're using green tea, which is more delicate than black or rooibos, cool the freshly boiled water for 5 minutes before pouring it over the tea.

Leave the tea to steep for 30 minutes or until fully cooled, then take out the tea bags or strain to remove tea leaves. Swirl in the honey. Pour into a large jar and add the jun kombucha scoby.

Cover the jar of honey with a clean cloth and secure with a rubber band. Leave to ferment at room temperature for at least a week or up to 2 weeks – the jun will become more sour, leaning towards apple cider vinegar in flavour, the longer you leave it.

Once you're happy with the flavour, within the 1–2 week window, pour the fermented tea through a funnel into a bottle. At this point you can add extra seasonal flavours (see below).

Secure the bottle with a lid and leave to ferment for a further day, at room temperature. During this second fermentation, the jun should become lightly fizzy – this happens because the bacteria in the drink will continue to feed from the honey and produce carbon dioxide, which builds up in the sealed bottle.

Once you're happy with the flavour and fizz levels, pop the jun in the fridge until you're ready to drink it. Open the lid on the bottle every couple of days to ensure too much carbon dioxide doesn't build up in the bottle. Your jun will keep nicely for 1–2 weeks but it will become tangier the longer you leave it.

Fruity variations

Rhubarb and rooibos Use rooibos tea. Add 1–2 rhubarb stalks to the jun once it is bottled, roughly chopping it enough to fit into the bottle. A few slivers of fresh ginger added to the mix is also delicious.

Plum and black tea Use a nice black tea or even classic builder's tea bags as the base. Once the jun is bottled, roughly chop 2 ripe plums and add them to the bottle (or purée the plums, strain through a sieve, then add the purée). For extra flavour, tuck in a sprig of lavender.

Apple and green tea Use your favourite green tea for the base. Once the jun is bottled, grate or chop 1 peeled and cored apple into pieces small enough to fit into the bottle, or juice the apple, then add. For an extra herbal layer, tuck in a few sprigs of lemon verbena or fresh mint.

Fennel honey G&T

Kat Jury, a friend and talented chef, dreamt up this elegant and refreshing tipple as the August sun set over River Cottage farmhouse. Honey and fennel are stunning together and marry well with the citrus and spice aromatics in gin.

Serves 1

1 tsp honey
A couple of sprigs of fennel fronds, with blossom if possible
A few ice cubes
50ml gin
200ml tonic

Drizzle a little of the honey on the inside of the glass. Put one of the fennel fronds into the glass with the ice cubes.

Put the remaining honey into a cocktail shaker with the gin, the other fennel sprig and a splash of the tonic water. Shake to mix, then strain into the honeyed glass. Top up with the remaining tonic water and serve.

Honey Mary with seaweed

Chilli heat and honey pair well and the combination gets even better when you shake sweet-sharp tomato juice and salty seaweed into the mix.

Serves 1

200ml tomato juice
50ml vodka
1 dried chipotle chilli, soaked in 50ml boiling water
1 tsp honey, plus a little for the glass
A few ice cubes
3 tbsp dried seaweed
A pinch of sea salt
A sprig of lovage, or a celery stalk, to garnish

Pour the tomato juice and vodka into a cocktail shaker, add the chilli with its soaking water, the honey and 2 tbsp of the seaweed.

Using a pestle and mortar, grind the remaining seaweed with the salt to a powder. Dip the tip of a spoon into honey and then run it around the rim of your glass to give it a really light coating. Now dip the honeyed rim into the seaweed salt to coat.

Shake the cocktail shaker to combine the mix. Pile the ice cubes into your glass and pour on the shaken Mary. Garnish with a little lovage or celery and enjoy.

Honeyed cocoa
with ginger and bay

This is really nourishing and as comforting as wrapping yourself in a woolly blanket in front of a crackling log fire. Ginger and bay both invigorate and soothe, while chocolate has reputed de-stressing attributes – it's high in magnesium which helps you to relax. So, if you're in need of a bit of TLC, this is a brilliant drink.

Serves 1

250ml milk (any kind – whole milk, nut milk or oat milk)
2 tbsp raw cacao or cocoa powder
2 slices of ginger, each 1cm thick
2 bay leaves
1–2 tsp honey, to taste

Pour half of the milk into a saucepan and warm gently. Add the cacao or cocoa and whisk until it is fully combined and there are no lumps. Add the remaining milk, the ginger and bay leaves. Heat gently for 5 minutes or until fully warmed through and the flavours of the ginger and bay are coming through.

You can strain the ginger and bay out or leave them in. Let the cocoa cool a little before sweetening with honey to taste, so the heat doesn't destroy its goodness.

Francesca's coffee

My Italian friend Francesca introduced me to the joys of adding honey to coffee. Its rounded kiss of nectar-rich sweetness is a beautiful coffee companion but it is also a healthy one, too. There are numerous studies showing that consuming a mug of honeyed coffee can soothe and even heal a persistent cough.

Serves 1

250ml freshly brewed coffee
1 tsp honey

Allow your coffee to cool slightly (or brew with cooled boiled water which almost always makes for a better pot of coffee); the cooler temperature preserves both the natural oils in the coffee beans and also the honey's goodness.

Swirl a teaspoonful of honey through your coffee, sip and enjoy.

Beauty and medicine

Making your own beauty products is the perfect extension to a 'from scratch' kitchen philosophy. As with food, when you make anything at home using pure ingredients you get the purest product but also one that nourishes more deeply. You can also harness the health-giving benefits of the hive's propolis by creating your own medicines.

Beeswax lip balm

This is a brilliant way of using beeswax and makes such a lovely gift. You can use all manner of different essential oils to flavour the balm – from Burt's-Bees-inspired peppermint oil, which offers a refreshing, cool soothe, to rose oil for an elegant, floral finish – or you can leave it natural.

Makes 90g

2 tbsp grated beeswax
2 tbsp shea butter
2 tbsp coconut oil
2–3 drops of essential oil, such as peppermint, rose or wild orange (optional)

You will also need

A few very small glass jars (or similar containers)

Put the grated beeswax, shea butter and coconut oil into a small heatproof bowl and set over a small pan of boiling water. Stir constantly until melted. Take the pan from the heat but keep the bowl over the still-hot water to keep the mixture melted.

Add your choice of essential oil, a few drops at a time, applying a tiny amount of the balm on your arm as you do so to test the fragrance and adjust to your liking.

Now transfer the lip balm to little jars for storage. It will set once cooled to room temperature.

Tip To clean the bowl after making the lip balm, while the residual wax is still warm, wipe with kitchen paper or a piece of fabric you can reuse for this purpose in future. Add 5 or 6 drops of essential lemon oil to the bowl and wipe to remove any excess. Clean with hot soapy water.

Honey moisturiser

Just the simple act of preparing this moisturiser is soothing, but you won't fully appreciate its luxurious qualities until you apply it to your skin.

Makes 200g

100ml cooled, boiled water
2 tbsp oats
4 tbsp olive or jojoba oil
4 tbsp coconut oil
2 tbsp grated beeswax
1 tsp raw honey (see p.144)
1 tsp vitamin E oil
2 drops of lavender essential oil

You will also need

A 200g lidded jar, or a bottle with a pump and a large piping bag

Put the water and oats into a bowl, cover and leave to soak for 12 hours (overnight works well).

Drain the oats in a sieve set over a jug or bowl to catch the liquid, which is what you'll use for the moisturiser. Press the oats gently with a spatula to extract all the oatmeal liquor; set this aside. Discard (or eat) the oats.

Put the olive or jojoba oil, coconut oil and beeswax into a heatproof bowl and set it over a pan one-third full of simmering water. Stir the mixture until just melted then take off the heat. Fold through the honey.

Put the oil mixture into the freezer for about 15 minutes or the fridge for around 30 minutes until solid.

Using an electric handheld or freestanding mixer, whisk the solidified mixture and oatmeal water together with the vitamin E oil and lavender essential oil until thickened and creamy.

Spoon into a jar or into a pastry bag to transfer the moisturiser to a bottle with a pump. Store at room temperature and use as needed.

Beeswax wraps

These are the perfect plastic-free solution for food storage. They look great and are easy to make with wax you've harvested from your hive or bought online, or obtained from a local beekeeper.

Makes 6–8

500g beeswax, fully cleaned, grated or cut into small 1cm pieces
1 metre 100 per cent cotton fabric

You will also need

Pinking shears or scissors
A paint or pastry brush (needs to be kept solely for this purpose)

Preheat the oven to 50°C/lowest gas setting. Cut the fabric into sizes that will fit on a large baking sheet. Use pinking shears to cut the fabric if you have them to help prevent fraying; otherwise scissors will do. For a snack bag, cut a 17 x 34cm piece of fabric. For a sandwich wrap, a 34cm square should be large enough.

Line a baking sheet with baking paper and lay the fabric on top. If your fabric is one-sided, place the patterned side face down. Use a fresh piece of paper each time you make another wrap.

Distribute a liberal amount of the beeswax evenly all over the fabric, making sure you get some pieces near the edges. Place in the oven for 4–8 minutes until the wax has melted completely. Take the tray out and use a paint brush to spread the wax evenly over the entire fabric. (The beeswax will stick to the brush, so you'll need to discard it after use, or save it to make future beeswax wraps.)

Using tongs, remove the fabric from the baking sheet; it should feel cool to the touch after waving it in the air for a few seconds. Hang the fabric up to dry or set it on the back of a chair, beeswax side facing up. Once the beeswax has set and is no longer very tacky, you can add buttons or hand-sew them into small pouches.

To make a snack bag, once dry, fold your beeswax-coated fabric in half with the non-treated sides facing inward. Sew the two side edges together, leaving the top open. Turn the bag inside out, and add a button as a closure or stitch Velcro to both top edges to fasten.

To make a sandwich wrap, sew a button in two adjacent corners on the patterned side of your wax-coated fabric square. To close, lay button-side down and fold the fabric into thirds around the sandwich. Flip and fold the ends of the fabric up so the buttons are on top. Wrap twine around for a secure closure.

To clean wraps after use, wash them by hand in cool water using a mild soap.

Beeswax candles

Opting for beeswax over soy and paraffin candles has countless benefits. Beeswax candles are non-toxic and environmentally friendly. They burn very cleanly with little smoke and help to neutralise pollutants in the air; this helps eliminate dust, odours and mould in the atmosphere, easing allergy and asthma symptoms and improving breathing for anyone nearby. They also have a high melting point (the highest among all known waxes), which results in a significantly longer burn time (2–5 times longer than soy and paraffin candles). And they drip very little, if at all.

Makes 6 jam jar candles

1kg beeswax
6 hemp or cotton candlewicks, each 18cm long and 1mm in diameter with sustainer tabs

You will also need

Six 200g jars
6 wooden lollypop sticks or chopsticks

Melt your beeswax using a double boiler (or in a heatproof bowl over a pan of gently simmering water), heating it to about 70°C. The great thing about beeswax is that you can re-melt it and start again, so if you need a bit of extra wax you can just melt more and top up your candles.

Have your jars and candlewicks at the ready. Make sure the wick will reach the bottom of the jar with enough extra to gently tie around a wooden stick, such as a lollypop stick or chopstick – this will hold the candlewick in place as the wax cools.

Once your wick is in place, pour the melted wax into the container. Leave to cool for at least 6 hours.

Remove the wick support stick from the top of the beeswax candle. Trim the wick, leaving about half of the excess wick sticking out of the top of the candle. Your beeswax candles are now ready to use.

Tip To clean the bowl after melting the beeswax, while the residual wax is still warm, wipe the bowl out thoroughly with kitchen paper or newspaper. Don't attempt to wash the bowl while it still contains some wax, as this will solidify and may block your sink.

Propolis tincture

Aristotle kept bees and showed a remarkably accurate and detailed knowledge of propolis. Ancient Roman naturalist and author Pliny the Elder, also a beekeeper, was a fan, too. To quote his *Naturalis Historia*, 'propolis is produced from the sweet gum of the vine or the poplar, and is of a denser consistency, the juices of flowers being added to it.'

Research confirms that this bee-gathered 'gum' or sap is rich with antibacterial, antifungal, antiviral and anti-inflammatory properties. It is said to protect the liver, to increase the body's natural resistance to viruses and other infections, to heal problems of the mouth and gums and so much more. To harness the potential benefits of this rich foodstuff, make this tincture and take a teaspoonful of it daily.

Makes 100ml

100ml vodka

2 tbsp propolis scrapings from the hive (or you can order them online)

You will also need

2 small lidded jam jars

Mix the vodka and propolis scrapings together in the jar. Put the lid on the jar and shake well. Store in a cool, dark cupboard and shake once a day for 2 weeks.

Strain the mixture through a muslin-lined sieve or paper coffee filter into a clean jar, seal and store in a dark place. You can collect and store the propolis left in the filter to use again.

Propolis throat spray

This powerful spray can inhibit bacterial throat infections such as strep throat. Spray in the back of the mouth anytime a sore throat hits.

Makes 90ml

3 tbsp propolis tincture (above)

2 tbsp honey (or elderberry tonic, p.227)

2–3 sprigs of thyme

1 tbsp cooled boiled water

You will also need

A small bottle with a spray top

Put all the ingredients into a spray bottle. Store in a dark cupboard at room temperature and use as needed. It will keep for 6 months.

Propolis-infused oil

Of all methods of infusion, research indicates that an oil extract of propolis may have the strongest antimicrobial effect. While you can apply this topically, as a soothing and healing oil for cuts, scrapes and dry skin disorders such as psoriasis or eczema, you can also use it as a flavoured oil for recipes. It has the most fantastic vanilla oaky smell with a hint of cedar and honey. I love making mayonnaise with it or simply using it to finish a dish like salt-baked celeriac.

Makes 250g

2 tbsp propolis scrapings from the hive (or you can order them online)
250ml olive or almond oil

You will also need
A 250ml bottle for storage

Mix the propolis and oil together in the top of a double boiler or a heatproof bowl set over a pan of just-boiled water. Heat the mixture to no higher than 50°C – use a thermometer to monitor the temperature (higher temperatures may destroy some of the beneficial properties contained in the propolis).

Stir and leave to infuse over the gentle heat for at least 30 minutes, up to 4 hours. The propolis won't completely dissolve but the warmth will help extract the flavour.

Strain the mixture through a muslin-lined sieve or a paper coffee filter into a jug. If you use muslin, you may have to filter the oil twice.

Store the finished oil in a sealed bottle in a dark place. The propolis remaining in the filter can be used again to make more oil – refrigerate or freeze for another time.

Useful Things

Directory

Bee friendly seeds

Sarah Raven
sarahraven.com
Extensive array of bee-loving flowers, available as plants, bulbs and seeds.

Tamar Organics
tamarorganics.co.uk
One of our favourite organic seed suppliers. The perfect starting point for creating a medicinal, bee-friendly herb garden.

Hives and supplies

Bee Kind Hives
beekindhives.uk
Go-to source for low-intervention bee hives such as Freedom and Warré hives. They also host log hive workshops.

National Bee Supplies
beekeeping.co.uk
Devon-based bee supplies with focus on sustainability, including National hives, made with sustainably sourced cedar. Also sell organic cotton bee suits, washable fabric and leather gloves, hive tools, starter kits and honey extractors.

Research

International Bee Research Association
ibra.org.uk
An international hive of information on bee science, encompassing all species, managed and wild. Excellent resource for seminal tomes, journals and papers on bees and honey.

Support

Blackbury Honey Farm
blackburyfarm.co.uk
Honey farm near River Cottage, rich with orchard and wild flower meadows. It offers extensive beekeeping courses and you can buy beeswax, honey and books in their shop or via their website.

British Bee Keeping Association
bbka.org.uk
An umbrella organisation for local beekeeping associations in Britain. It will help you find one near you that will give you support from experienced local beekeepers, as well as access to training and equipment.

Garden Organic
gardenorganic.org.uk
Invaluable resource supporting organic gardening with rare 'bee-friendly' seeds through their Heritage Seed Library.

National Bee Unit (aka BeeBase)
nationalbeeunit.com
The organisation to register your hives if you are keeping bees. It helps provide information about local threats to bees, such as Yellow-legged/Asian hornet sightings or the spread of disease.

Natural Bee Keeping Trust
naturalbeekeepingtrust.org
A support system for beekeepers with low-interventions hives, such Warré, Top bar or Freedom (Rocket). Also a valuable source of information and advice if you're considering beekeeping but aren't sure which approach to take.

Acknowledgements

First and foremost, thanks and praise to the River Cottage bee guardian and co-author of this book, Steve Minshall. Not only does Steve do a magnificent job caring for the River Cottage honeybees, but he also brings invaluable knowledge and experience across the pollinator world. In addition, he is responsible for most of the up-close-and-personal insect and bee colony images through the book.

Further tributes are due to David Chambers, who helped establish the first River Cottage apiaries, and Matt Somerville, our 'bee whisperer', who helped us to create more natural bee habitats and bee-friendly hives.

To Hugh Fearnley-Whittingstall, a huge thank you. Without Hugh, these informative handbooks would not exist and he's been fully involved in this book from the start. I'm forever grateful to Stewart Dodd for his patience and support, and to Antony Topping of Greene & Heaton for ensuring this book happened!

Sarah Wyndham Lewis and Dale Gibson of Bermondsey Street Bees, held in high esteem for their ethical beekeeping, have been beyond generous with their knowledge. And Petersham Nurseries in London have always supported my River Cottage projects. Thanks also to our local family-run Blackbury Honey Farm set up by Ken, Maureen and Daniel Basterfield, who top up our honey supply and have a beautiful apiary. And to the East Devon Beekeepers for all their support, with special mentions for Richard Simpson, Keith Bone and Val Bone.

Photographer Ali Allen and her sons Wilf, Bertie and Boo are woven into the fabric of this book. We had lots of adventures along the way – including a few stings and lots of honey cake! Thanks to Theodora Cross-Dodd for her vital contributions, to Kat Padgett who created a few recipes, and to my son Rory Gibson who helped with lots of research and writing around plants for pollinators. I'm also grateful to the Rebel Family in Montes de Málaga for providing honey, inspiration and the perfect writing space during our stay with you.

Helen Jukes, author of *A Honeybee Heart Has Five Chambers*, was tremendously helpful and generous with her time and knowledge. Similarly, I'm incredibly grateful to Robin Snowdon, Tara King, Brett, Nici and Bronte Cooper, Josh Pollen, Mike Knowlden and Mark McCabe, who have all been brilliant.

Last but not least, the vital worker bees who I've had the pleasure to work with on several projects. Will Webb not only beautifully designed this book but, as a beekeeper himself, helped frame the book. Janet Illsley, thank you, it's always a great pleasure to work with you. Sally Somers, with your beekeeping experience, your final read has been invaluable. At Bloomsbury, thanks to original editor Kitty Stogdon, and Samhita Foria who took it over the finish line; and to Rowan Yapp who brings authors' ideas to life and more beautiful, informative books to the world.

As for the bees who keep our landscape flourishing… let's look after them.

Rachel

Index

Page numbers in *italic* refer to the illustrations

abdomen, anatomy 60
acarine mites 20
aggressive bees 114
allergy, to stings 117
American foulbrood (AFB) 118
anaphylactic shock 117
anatomy 58–60, *59*
angelica 45
antihistamine tablets 90
apiary sites 92–4
apple blossom honey 148
apple cider vinegar: honey herb-infused vinegar *172*, 173
apples: apple and green tea 231
 apple baklava 209–11, *210*
arbutus honey 149
ashy mining bee 30–1, *30*
Asian hornet 121–2, *121*
aubergine jackets 188
avocado honey *146*, 150
Ayurvedic medicine 142

baklava, apple 209–11, *210*
bananas 116
 banana mousse with honey *200*, 201
 honey tahini flapjacks *220*, 221
basil 46, 173
bay 173
beans, honey 184, *185*
beauty products 236–9
bee bread 64, 68, 143
bee brushes 89
bee flies 32, *33*
bee hotels 37
Bee Kind Hives 80
bee space 78–9, 86
bee suits 116
bees: aggressive bees 114
 anatomy 58–60, *59*
 castes and behaviours *59*, 61
 development *100*
 evolution 13–14
 habitats 37–9
 pollination 11, 13, 14
 swarms 73, 78, 88, 106–9, *107*
 types of 11, 20–32
 see also drones; workers; queens
beeswax 77, 140, 163
 capped cells 141, *162*, 163
 foundation 86, 87
 harvesting 152, 163
 production of 62
beeswax candles 243
beeswax-fermented fruit *222*, 223
beeswax lip balm 236
beeswax wraps *240*, 241
behaviour 61–7
bergamot 46
black bees 20
blackberry and honey cream *192*, 193
boots 89, 116
borage 46, *47*, 66, 148
bottling honey 158–9
Box Tree hive 80, *81*
bread: honeyed summer pudding *190*, 191
British Beekeepers Association (BBKA) 77, 94, 156
British Isles 20–32
brood boxes 85, 90, 103, *137*
brood pattern 104
brown carder bee 24–5
buckets 90, 156
Buckfast bee 20, 21, *21*
buckwheat honey *146*, 150
buff-tailed bumble bee 23
bumble bees 21–5, 39
buns, honey *218*, 219

cage bottles 91
cake, honeyed carrot 216, *217*
candles, beeswax 243
capped cells 141, 152, *153*, 157–8, *162*, 163
carder bees 24–5, *24*
Carr, William Broughton 79
carrot cake, honeyed 216, *217*
cast swarms 108
castes *59*, 61
cave paintings 15, 78
cells: bee development *100*
 brood boxes 85
 capped cells 141, 152, *153*, 157–8, *162*, 163
 queen cells 71, *72*, 73, 104, 108, 109–11, *109*, 115
chalkbrood 123–4
chamomile honey jelly 194, *195*
cheese: feta-stuffed figs with honey 178, *179*
chestnut honey 150
chillies: sweet chilli sauce *170*, 171
chives 44, 46
chocolate: chocolate truffles 198
 honeyed chocolate pots 198, *199*
 honeyed cocoa 234, *235*
Christmas quincemeat 224, *225*
classifications, honey 151
clearer boards 91
clothing, protective 89, 91
clover 148
coffee, Francesca's 234
colonies *see* hives
Colony Collapse Disorder (CCD) 11, 12

colour, flowers 66
comb: brood boxes 85
building 62–4, *63*, *64*
capped cells 141, 152, *153*, 157–8, *162*, 163
drone eggs 68, *69*
early hives 78
egg laying in 72–3
foundation 86
harvesting honey 154
heather honey 160
hive notes 105
nectar storage 140
queen cells 71, *72*, 73
River Cottage hives 80
supers 87
Top Bar hives 82, *82*
Warré hives 83, *83*
comb honey 151
comfrey 47
common carder bee 24, *24*
common mourning bee 27
communication, dance 66–7, *67*
containers, extracting honey 156
coriander 47–8, *146*, 150
cotton blossom honey *146*, 148
cream: blackberry and honey cream *192*, 193
honey custard *204*, 205
honeyed chocolate pots 198, *199*
honeyed hop panna cotta *196*, 197
crown boards *84*, 87, 101, 102, 160
custard, honey *204*, 205
custard tart, roasted squash *206*, 207–8

dances, communication 66–7, *67*
dandelion honey 149
Dark British bee 20
dark-edged bee fly *33*
dates: honey tahini flapjacks *220*, 221
honeyed carrot cake 216, *217*
roasted squash custard tart *206*, 207–8
deformed wing virus (DWV) 123, *123*
dill 48
diseases 63, 105, 118–25
drained honey 151
dressing, smoky honey 180
drinks: fennel honey G&T 232, *233*
Francesca's coffee 234
honey Mary with seaweed 232, *233*
honeyed cocoa 234, *235*
jun 230–1
mead *228*, 229–30
drones 37–9, 58, *59*, 61, 68–9, *69*, *100*
dukka, Hugh's honey and *176*, 177
dummy boards 86, 101, 102
Dzierzon, Jan 78

early bumble bee 23–4, *23*
echinacea 44, 48, *49*
eggs 14, *63*, *100*
brood boxes 85
drones 68, 69
laying 71, 72–3
not seen on inspection 111
queen excluders 87
setting up a hive 96
solitary bees 40
worker bees 61
ekes 90, 96, 163
elderberry tonic *226*, 227
epigenetics 71
equipment: extracting honey 156–7
National hives 88–91
European foulbrood (EFB) 118–19
extinction of bees 11
extracted honey 151
extracting honey 90, 152–60, *155*, *159*
eyes 58, *59*, 66

Fearnley-Whittingstall, Hugh 80
feeders 89, 96, 115–16
fennel 48
fennel honey G&T 232, *233*
figs: feta-stuffed figs with honey 178, *179*
filtered honey 151
filters 159–60
flapjacks, honey tahini *220*, 221
flight paths 93
floral honeys 148
flower bee, hairy-footed 26, *26*
flowering plants 42–55, 93
and flavour of honey 147–50
foraging 66–7
nectar 140
pollination 13–14
fondant 91
foraging 65, 66–7
foulbrood 118–19
foundation, beeswax 86, 87
frames *84*, 86
extracting honey 156–7
inspecting hives 101–2
setting up a hive 95, 96
spacing 78–9
supers 87
Francesca's coffee 234
Freedom hive 16, *79*, 80, *81*
fruit: beeswax-fermented fruit *222*, 223
honeyed summer pudding *190*, 191
fuel, smokers 89

gardens 36, 42–55
genetics 61, 68, 71

gin: fennel honey G&T 232, *233*
ginger honey teriyaki 183
gloves 89, 116, *117*
Gooden's nomad bee 28, *29*
gooseberries, sardines with *182*, 183
green-eyed flower bee 27
grot boxes 90
guarding hives 64–5, *65*, 99

habitats 37–40
hairy-footed flower bee 26, *26*
halva, honey *214*, 215
haricot beans: honey beans 184, *185*
harvesting: beeswax 152, 163
  honey 152–60, *155*, *159*
  pollen 165
  propolis 152, 164–5
hawthorn honey 149
hazelnuts: dukka *176*, 177
head, anatomy 58–60, *59*
healing foods 45
heather honey *146*, 147–8, 150, 160
hedgerows 36
herbs 44–51
hive floors 85
hive notes 102, 104–5, *105*
hive stands 88, 89, 93, 95
hive tools 89
hives 78–91
  acquiring bees 88
  apiary sites 92–4
  bee space 78–9
  the beekeeper's year 128–36
  cleaning supers 160
  frames 78–9
  guarding 64–5, *65*, 99
  history 15–16, 78–9
  hygiene 63
  inspecting 96–105, *98*, 111–12
  insufficient stores 115–16
  introducing new queens 114–15, *114*
  merging 111, 112–13, *113*
  overcrowding 110
  queen-less hives 112
  setting up 94–6
  splitting 110–11
  swarming 106–8, *107*
  types of 16, 80–8
  weak hives 110, 113
Hoffman frames 79, 85, 86
honey: bottling 158–9
  classifications 151
  cooking with 168–9
  crystallisation 160
  flavours 147–50
  harvesting 14–15, 90, 152–60, *155*, *159*
  honey laundering 144, 147
  medicinal properties 141–2
  pollen in 143
  production of 140–1, 152
  raw honey 144–5
  ripeness 152
  stores in hive 116, 152
  vs sugar 142–3
honey beans 184, *185*
honey buns *218*, 219
honey custard *204*, 205
honey-glazed pork belly 189
honey halva *214*, 215
honey ice cream 205
honey-infused vinegar *172*, 173
honey Mary with seaweed 232, *233*
honey moisturiser *238*, 239
honey presses 160
honey roast pears 202, *203*
honey stomach 65
honey tahini flapjacks *220*, 221
honeycomb *see* comb
honeydew honey 151
honeyed carrot cake 216, *217*
honeyed chocolate pots 198, *199*
honeyed cocoa 234, *235*
honeyed hop panna cotta *196*, 197
honeyed summer pudding *190*, 191
hops: honeyed hop panna cotta *196*, 197
hornets 32, *33*, 121–2, *121*
hoverflies 32, *33*
Hugh's honey and dukka *176*, 177
hygiene 154
hyssop 48

ice cream, honey 205
inspecting hives 96–105, *98*, 111–12
introduction cages 114–15, *114*
Isle of Wight disease 20
ivy mining bee 31, *31*

jars 90, 158–9, 161
jelly, chamomile honey 194, *195*
jun 230–1

labelling 144, 147, 161
landing boards 88
Langstroth hives 78–9
larvae 14, 40, 61–2, *63*, 68, 71, 87, *100*, 123
lavender 44, 50, 66, 161
leaf-cutter bee 27
lemon honey sorbet 212, *213*
lemon balm 50
lemon verbena 44, 173
lifespans 37, 68, 69
ling honey *146*, 150

lip balm, beeswax 236
logbooks 90
long-horned bee 27
lovage 173

manuka honey 142, 147, 150
marigolds *52*
marking queens 91, 97, *97*
masked bee 28
mason bees 28–9
mating 68–9, 71
mead *228*, 229–30
medicinal properties, honey 141–2
mentors 94
mice 125
milk: honeyed cocoa 234, *235*
minerals 141, 143, 168
mining bees 30–1, 40
mint 50–1, 173
miso, spiced honey 188
mobile phones 90
moisturiser, honey *238*, 239
moths 124, *124*
mourning bee, common 27
mousse, banana with honey *200*, 201

Nasonov's glands 64, 96, 109
National Bee Unit (NBU) 118
National hives 16, 79, *84*, 85, 88–91, 95, 131–6
nectar 13, 36, 40–4, 63, 65, 140
neonicotinoids (neonics) 40–1, 145
nests 14, 37, *38*, 39–40
nosema 125
nucleus, setting up a hive 94, 95–6
nucleus boxes 90, 108, 110
nuts: apple baklava 209–11, *210*

oak honey 150
oats: honey tahini flapjacks *220*, 221
oil, propolis-infused *246*, 247
oilseed rape honey 148, 160
orange, shallot tatin with 186–7, *187*
orange blossom honey 149
orange-tailed mining bee 31
orange-vented mason bee 29
oregano 50, 173
organic honey 42, 44, 145, 151
overcrowded hives 110

panna cotta, honeyed hop *196*, 197
pastries: apple baklava 209–11, *210*
pastry: rough puff pastry 186–7
  shortcrust pastry 207–8
pears, honey roast 202, *203*
pesticides 40–2, 145
pests 118–25
pheromones 64, 68, 71, 73, 99, 109, 112, 115
plum and black tea 231
pollen 64, *64*, 88–91, 95, 104–5, 141, 143, 151
pollen baskets 66, 164
pollen traps 165
pollination 11, 13, 14, 36, 41
pork belly, honey-glazed 189
Porter bee escapes 87, 91, 160
pressed honey 151
propolis 66, 140, 152, 164–5
  propolis-infused oil *246*, 247
  propolis throat spray 244
  propolis tincture 244
protective clothing 89, 91

queen cells 71, *72*, 73, 104, 108, 109–11, *109*, 115
queen excluders *84*, 86–7, 95, 103, 112–13
queens 14, 37–9, 58, *59*, 61, *70*, 71–3
  aggressive bees 114
  brood boxes 85
  bumble bees 39
  creation of 71, 73, *100*
  death 73, 109
  egg laying 72–3
  inspecting hives 97, 102–3, 104
  introducing new queens 114–15, *114*
  marking 91, 97, *97*
  mating 68–9, 71
  queen cannot be found 112
  queen-less hives 112
  retinue 72
  setting up a hive 96
  splitting hives 110–11
  swarming 73, 106–9
  in winter 73
quinces: Christmas quincemeat 224, *225*
  quince baklava 211

rata honey 149
raw honey 144–5
red mason bee 29, *29*
red-tailed bumble bee 22, *22*
refractometers 91, 156
rhubarb: honey-glazed pork belly 189
  rhubarb and rooibos 231
Rocket hive 80, *81*, 91
roofs 88, 101
rooibos, rhubarb and 231
rosemary 51, 66, 149, 173
rough puff pastry 186–7
royal jelly 61, 62, 68, 71, 165

sacbrood 123
sage 51, 173
salad, summer tomato 180, *181*
sardines with gooseberries *182*, 183

sauces: ginger honey teriyaki 183
sweet chilli sauce *170*, 171
seeds, gardening 44
shallot tatin with orange 186–7, *187*
shortcrust pastry 207–8
shrill carder bee 25
Sidr honey 150
sieves 158, *159*
siting hives 92–4
skeps 78, 108
small hive beetle 119
small scissor bee 28
smokers 89, 91, 99
soil 36
solitary bees 14, 20, 26–8, 40
Somerville, Matt 80
sorbet, lemon honey 212, *213*
sperm 61, 68–9
squash: roasted squash custard tart *206*, 207–8
sterilising jars 158
stingers and stings 58, 60, 65, 71, 116–17, *117*
strainers 90, 156
strawberries: beeswax-fermented fruit *222*, 223
sugar syrup, feeding colonies 115–16
sugar vs honey 142–3
summer pudding, honeyed *190*, 191
summer savory 173
summer tomato salad 180, *181*
sunflower honey *146*, 149
supers *84*, 87, 103, 105, 160
supersedure queen cells 110
swarms 73, 78, 88, 106–9, *107*
sweat bee 28
sweet chilli sauce fermented with honey *170*, 171
symbolism 19

tahini: honey halva *214*, 215
honey tahini flapjacks *220*, 221
tarragon 173
tarts: roasted squash custard tart *206*, 207–8
shallot tatin with orange 186–7, *187*
tawny mining bee 30
tea: apple and green tea 231
jun 230–1
plum and black tea 231
rhubarb and rooibos 231
thorax, anatomy 60
throat spray, propolis 244
thyme 44, 45, 51, 149, 173
tomatoes: honey Mary with seaweed 232, *233*
summer tomato salad 180, *181*
tonic, elderberry *226*, 227
Top Bar hives 80, *81–2*, 82, 91, 132–4, 160
travel screens 91
tree bumble bee 25, *25*
tropilaelaps mite 119
truffles, chocolate 198

ultraviolet light 44, 66
uncapping boxes 91
uncapping tools 91

varroa 16, *69*, 119–20, *120*
and deformed wing virus 123
hive floors 85
treatments 45, 51, 91, 120, *137*
veils 89, 116
venom sacs 116, *117*
Verbena bonariensis 43
Veterinary Medicines Directorate (VMD) 120
vinegar, homemade honey 174, *175*
vitamins 141, 143, 168
vodka: honey Mary with seaweed 232, *233*
propolis tincture 244

Warré hives *17*, 80, *81*, 83, *83*, 91, 132, 134, 152, 160
wasps 99, 125
water 65, 157
wax *see* beeswax
wax moths 124, *124*
WBC hives 79
weak colonies 110, 113
weather, hive notes 105
wellington boots 89, 116
white-tailed bumble bee 22–3, *22*
wild bees 37–9
wildflower meadows 36
woodpeckers, green 125
wool carder bee 27
workers 14, 37–9, 58, *59*
beeswax production 163
creation of new queens 73
dances 66–7, *67*
development *100*
egg-laying 69
foraging 65–6
lifespan 37, 68
population in hive 68
production of honey 140–1, 152
queen's retinue 72
sensory capabilities 66
swarming 73, 106–8
tasks in hive 61–5
in winter 39, 73
wraps, beeswax *240*, 241

yellow-faced bee 28
yellow-legged hornet 121–2, *121*
yoghurt: honeyed carrot cake 216, *217*
honeyed hop panna cotta *196*, 197

River Cottage Handbooks

Seasonal, Local, Organic, Wild

FOR FURTHER INFORMATION AND
TO ORDER ONLINE, VISIT
RIVERCOTTAGE.NET